T0313870

Autonomous Road Vehicle Path Planning and Tracking Control

Autonomous Road Vehicle Path Planning and Tracking Control

Levent Güvenç
Bilin Aksun-Güvenç
Sheng Zhu
Şükrü Yaren Gelbal

IEEE Press Series on Control Systems Theory and Applications
Maria Domenica Di Benedetto, Series Editor

IEEE PRESS

WILEY

Library of Congress Cataloging-in-Publication Data:

Names: Güvenç, Levent, author. | Aksun-Güvenç, Bilin, author. | Zhu,
   Sheng author. | Gelbal, Şükrü Yaren, author. |
   John Wiley & Sons, publisher.
Title: Autonomous road vehicle path planning and tracking control / Levent
   Güvenç, Bilin Aksun-Güvenç, Sheng Zhu, Şükrü Yaren Gelbal.
Description: Hoboken, New Jersey : Wiley-IEEE Press, [2022] | Includes
   bibliographical references and index.
Identifiers: LCCN 2021047513 (print) | LCCN 2021047514 (ebook) | ISBN
   9781119747949 (cloth) | ISBN 9781119747956 (adobe pdf) | ISBN
   9781119747963 (epub)
Subjects: LCSH: Automated vehicles–Design and construction. | Automated
   vehicles–Collision avoidance systems. | Mathematical
   optimization–Industrial applications.
Classification: LCC TL152.8 .G88 2022 (print) | LCC TL152.8 (ebook) | DDC
   629.04/6–dc23/eng/20211101
LC record available at https://lccn.loc.gov/2021047513
LC ebook record available at https://lccn.loc.gov/2021047514

Cover Design: Wiley
Cover Images: Background image © Pobytov/Getty Images,
© Gorodenkoff/Shutterstock

Levent Güvenç and Bilin Aksun-Güvenç dedicate this book to their parents and to their son Kunter Güvenç.
Sheng Zhu dedicates this book to his beloved family.
Şükrü Yaren Gelbal dedicates this book to his beloved family.

Contents

About the Authors

Levent Güvenç is a professor of mechanical and aerospace engineering at the Ohio State University (OSU) with a joint appointment at the electrical and computer engineering department. He is a member of the International Federation of Automatic Control (IFAC) Technical Committees on Automotive Control; Mechatronics; and Intelligent Autonomous Vehicles and the IEEE Technical Committees on Automotive Control (past member); and Intelligent Vehicular Systems and Control (founding member). During 1996–2011, Levent worked in the mechanical engineering department of Istanbul Technical University where he was the director of the European Union Framework Research Programme 6 funded Center of Excellence on Automotive  Control and Mechatronics. He served as department chair of mechanical engineering at Istanbul Okan University from 2011 to 2014, was the founder and director of Mekar Mechatronics Research Labs and coordinator of team Mekar in the 2011 Grand Cooperative Driving Challenge. He is the co-founder and co-director of the Automated Driving Lab at the Ohio State University. He is the co-author of 1 edited volume, 3 books including this one, 4 book chapters, and more than 220 publications in major journals and conference proceedings. He is an ASME fellow. He was the chair of the 2007 IEEE Intelligent Vehicles Symposium and has been a regular program committee member of IEEE Intelligent Vehicles and Intelligent Transportation Systems conferences. Levent has been working on ADAS and connected and automated driving system development with the automotive industry for more than two decades. He is the co-inventor of six patents including two European patents, developed jointly with the automotive industry. He also has two provisional patent applications at OSU. His work on connected and autonomous driving has resulted in nine research prototype vehicles (two of them with a major automotive OEM) with different levels of autonomy, the last four being in the Automated Driving Lab at OSU. Levent was a member of the USDOT Smart Columbus project workgroups on Autonomous Electric Vehicles and the Connected Vehicle Environment.

Bilin Aksun-Güvenç is a Research Professor in the Department of Mechanical and Aerospace Engineering of the Ohio State University since September 2017. She joined the Ohio State University in September 2014 as a Visiting Professor in the Department of Mechanical and Aerospace Engineering after working in Istanbul Technical University (17 years) and Istanbul Okan University (3 years) as a professor. Her expertise is in automotive control systems – primarily autonomous vehicles, connected autonomous vehicles, adaptive cruise control, cooperative adaptive cruise control, collision avoidance systems, cooperative collision avoidance systems, electronic stability control, vehicle dynamics controllers, intelligent transportation systems, smart mobility, and smart cities. She has a long history of working as a principal investigator and researcher in several automotive industry projects for major automotive OEMs. She is the co-inventor of two European patents on yaw dynamics stability control and yaw rate virtual sensing for road vehicles, respectively. She is also the co-inventor of two provisional patent applications on autonomous vehicle safety testing and autonomous driving function evaluation and development. She is a member of the IFAC technical committees on Automotive Control (since 2005) and on Mechatronics (since 2008). She is the author of 2 books including this one, 2 book chapters, and over 130 publications in technical journals and conferences. Prof. Bilin Aksun-Güvenç participated in the NIST Global Cities Technical Challenge technical clusters Smart Shuttle and SMOOTH with the City of Columbus. Prof. Aksun-Güvenç led the Connected Electric Vehicle Deployment project which was part of the cost share contribution of the Ohio State University to the Smart Columbus project. The project developed pre-deployment MIL and HIL simulation evaluation of autonomous shuttle operation in the Linden Residential Area before the actual deployment was allowed to start. She is the co-founder and co-director of the Automated Driving Lab at the Ohio State University.

Sheng Zhu received the BS and MS degrees in automotive engineering from Beijing Institute of Technology and Tongji University, China, in 2012 and 2015, respectively. He received his PhD in 2020 in mechanical engineering from the Ohio State University, Ohio, USA. His research interests include on-road trajectory planning and robust control with HIL and experiment applications in automated vehicles under all-weather conditions, with over 10 research articles published and presented in peer-reviewed journal and conferences. Dr. Zhu's honors include recipient of National Scholarship of China in 2009, and graduation with the highest GPA in automotive engineering, Beijing Institute of Technology, 2012. Dr. Zhu currently works as a software engineer on planning and control at DeepRoute.ai in developing level-4 autonomous driving technology, with project development experience including four-wheel-driven vehicle control, autonomous valet parking, low-level torque steer control, online calibration system, weight estimation and compensation, etc.

Şükrü Yaren Gelbal is a PhD student in the Ohio State University Electrical and Computer Engineering Department and a Graduate Research Associate in the Automated Driving Lab. Before joining the Ohio State University as a visiting scholar then pursuing his PhD, he received the BS degree in mechatronics engineering from Okan University, Istanbul, Turkey in 2014 and MS degree in mechatronics engineering from Istanbul Technical University, Istanbul, Turkey in 2017. During his studies, he worked in numerous industry-funded, government-funded, and non-funded research projects while authoring and co-authoring more than 30 papers with overall more than 150 citations. From 2013 to 2014, he was a research intern with the Automotive Controls and Mechatronics Research Lab in Okan University. From 2015 to 2016, he was a researcher at Istanbul Technical University for a project which was financially supported by the Scientific and Technological Research Council of Turkey. His research experience includes software development, software and hardware implementation for CAVs and UAVs, HIL simulations, collision avoidance, V2X communication, and VRU safety. Mr. Gelbal's honors include researcher scholarship by Scientific and Technological Projects Funding Program supported by the Scientific and Technological Research Council of Turkey for Master of Science students, 2015, graduation with High Honors from mechatronics engineering, Okan University, 2014, graduation with the Highest GPA from mechatronics engineering, Okan University, 2014.

Preface

Autonomous driving of road vehicles is an important topic with an increasing number of successful real-world applications and the expected introduction of series production vehicles with Society of Automotive Engineers (SAE) Level L3 and SAE Level L4 autonomous driving capabilities by automotive companies around the world into the market. This trend has considerably increased both academic and industry research in this area.

The most fundamental task of an autonomous vehicle is the ability to plan and follow a path while avoiding collisions. Uncertainties in the environmental conditions, vehicle dynamics, vehicle load, and load distribution and the range of required speed from very low speeds to highway driving speeds require the path tracking and collision mitigation controls to be robust. This book contributes to this area by presenting the recent research results of the authors in path planning and robust path-tracking control. The methods presented in the book are applicable in real life, having been tested in a realistic hardware-in-the-loop simulation environment and in road testing with a research-level autonomous vehicle along with the usual model-in-the-loop simulations.

The target audience for this book is both researchers and practitioners working on autonomous vehicle motion planning and control, with the main target community being control scientists and engineers working on autonomous driving. This book focuses on the applications of robust motion control to the autonomous driving part of the automotive area. Students, especially graduate students, in this research area will also find the book of interest to them. Chapters 1, 2, 4, and 5 are based partially on material taught in courses by Prof. Levent Güvenç and Prof. Bilin Aksun-Güvenç. These include the Ohio State University courses ECE 5553 Autonomy in Vehicles (graduate and senior undergraduate) and ME 8322 Vehicle System Dynamics and Control (graduate) and the past Istanbul Technical University and Istanbul Okan University courses on Automotive Control Systems (graduate) and Vehicle Dynamics and Control (undergraduate). The doctoral research work of Drs. Sheng Zhu and of Şükrü Yaren Gelbal (in progress) at the Ohio State university have also contributed significantly to this book.

The authors' graduate students and research collaborators have also made contributions to several topics covered in the book through joint publications and their names may be found in the references cited at the end of each chapter.

We hope that this book will be a useful reference for students, researchers, and practitioners interested in or working in the autonomous vehicle path planning, path-tracking, and collision avoidance areas.

Columbus, OH, USA
2021

Levent Güvenç
Bilin Aksun-Güvenç
Sheng Zhu
Şükrü Yaren Gelbal

List of Abbreviations

ABS	anti-lock brakes
ACC	adaptive cruise control
ADAS	advanced driver assistance systems
AEB	automatic emergency braking
API	application programming interface
AV	autonomous vehicle
BFGS	Broyden–Fletcher–Goldfarb–Shanno
BNP	Bakker–Nyborg–Pacejka
BSM	basic safety message
CALTRANS	California Department of Transportation
CAN	controller area network
CC	cruise control
CDOB	communication disturbance observer
CMU	Carnegie Mellon University
CNN	convolutional neural network
CPU	central processing unit
CRB	complex root boundary
CV	connected vehicle
DDOB	double disturbance observer
DLC	double-lane change
DOB	disturbance observer
DOF	degrees of freedom
DP	dynamic programming
DYC	direct yaw-moment control
ESC	electronic stability control
ESP	electronic stability program
FRM	frequency response magnitude
FSM	finite state machine
FWS	front wheel steering
GCDC	grand cooperative driving challenge
GPS	global positioning system
GPU	graphical processing unit
HAD	highly automated driving
HIL	hardware-in-the-loop
IMU	inertial measurement unit

IRB	infinite root boundary
ITS	intelligent transportation systems
JOSM	Java OpenStreetMap
KKT	Karush–Kuhn–Tucker
LCAS	lane change assistance system
LIDAR	laser imaging, detection, and ranging
LKAS	lane keeping assistance system
LKS	lane keeping system
LMI	linear-matrix-inequality
LPV	linear parameter-varying
MABx	dSPACE MicroAutoBox
MF	magic formula
MIL	model-in-the-loop
MPC	model predictive control
MSTE	mean squared tracking error
NAHSC	National Automated Highway System Consortium
NHTSA	National Highway Traffic Safety Administration
NLOS	non-line-of-sight
NPC	non-player character
OBU	on-board unit
ORU	other road users
OS	overshoot
OSM	OpenStreetMap
P	proportional controller
PATH	California partners for advanced transit and highways
PCL	point cloud library
PD	proportional-derivative controller
QP	quadratic programming
RALPH	rapidly adapting lateral position handler
RCA	Radio Corporation of America
RGB	red-green-blue
RLS	recursive least squares
RMS	root mean square
ROS	robot operating system
RRB	real root boundary
RRT	rapidly exploring random-trees
RSU	roadside unit
RTK	real-time kinematics
RWS	rear wheel steering
SAE	Society of Automotive Engineers
SCMS	security credential management system
SIL	software-in-the-loop
SISO	single-input single-output
SLAM	simultaneous localization and mapping
SOTIF	safety of the intended functionality
SQP	sequential quadratic programming
SRI	Artificial Intelligence Center of Stanford Research Institute

SVM	support vector machine
TCS	traction control system
V2X	vehicle-to-everything
VRU	vulnerable road user
VSC	vehicle stability control
VVE	vehicle-in-virtual-environment
XIL	X-in-the-loop

1

Introduction

This chapter provides an introduction to the whole book. After a section on motivation and introduction, a brief history of automated driving is presented, followed by how Advanced Driver Assistance Systems (ADAS) naturally evolved into autonomous driving functions. Some past and current autonomous driving architectures are presented using examples from the field. A literature review section where the key papers and more recent developments in path planning and robust path-tracking control for autonomous road vehicles, also including the relevant literature on cybersecurity, and how it relates to autonomous vehicle path planning and tracking, are summarized next. This is followed by a section on the scope of the book, briefly detailing what is covered in each chapter. The chapter ends with a brief summary and concluding remarks.

1.1 Motivation and Introduction

The race toward series produced autonomous road vehicles has been rapidly progressing during the last decade. Most automotive OEMs and technology companies had promised or forecasted autonomous driving models by the year 2020, two years before the publication date of this book. This obviously did not take place. While we do not have truly autonomous driving vehicles that the public can currently buy, the currently available lane keeping, adaptive cruise control (ACC), emergency braking systems, traffic jam assistants, and their extended versions in some vehicles allow an almost autonomous highway driving experience under ideal conditions [1]. Autonomous shuttle service has been successfully deployed in a lot of different geofenced areas worldwide [2–4]. Large-scale autonomous taxi service is about to start in several countries in Asia soon, using drive-by-wire vehicles retrofitted with sensors and control systems [5]. Autonomous vehicle races have also been increasing around the world [6]. Autonomous delivery vehicles and autonomous truck platoons are also technologies with many successful, limited-scale deployments [7, 8]. Automotive OEMs were planning to introduce autonomous products for the fleet market first, before making them available to the general public. Introduction of autonomous vehicle fleets that can also be used as ride hailed taxis is now expected by the year 2023 even though there may still be delays considering the failed predictions of the recent past. The current technology of traditional and nontraditional automotive OEMs and technology companies like Google's Waymo, and similar ones is sufficiently advanced for nearly full driverless operation in well-mapped environments under ideal conditions. The relatively smaller percentage of nonideal conditions and uncertain environments make it difficult to implement full-scale autonomous driving under arbitrary conditions and environments.

Autonomous Road Vehicle Path Planning and Tracking Control, First Edition.
Levent Güvenç, Bilin Aksun-Güvenç, Sheng Zhu, and Şükrü Yaren Gelbal.
© 2022 The Institute of Electrical and Electronics Engineers, Inc. Published 2022 by John Wiley & Sons, Inc.

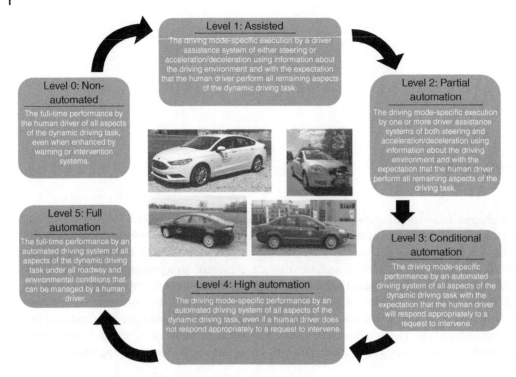

Figure 1.1 Categories of autonomous driving according to SAE.

Autonomous road vehicles have been categorized into six categories by the Society of Automotive Engineers (SAE) as shown in Figure 1.1 [9]. Currently available automated driving technology in series produced vehicles falls under Level 2 which is partial automation. Level 2 partial automation is achieved in series production vehicles with lane-centering control for steering automation and ACC and collision avoidance for automation in the longitudinal direction. L3 partial automation is characterized by all driving actuators being automated and the presence of a driver who can intervene when necessary. Recently introduced autopilot systems for cars are examples of conditional automation where the car takes care of driving in some driving modes like highway driving but the human operator is always in the driver seat to take over control if necessary. The Highway Chauffeur is a Level 3 autonomous highway driving system in which almost all highway driving functions are carried out autonomously, but the driver is needed to take over if something goes wrong or might go wrong like a lane change maneuver [10]. The Highway Chauffeur is currently available technology for series produced vehicles and uses an eHorizon electronic map to take care of driving on the highway until the chosen exit is reached. The Highway Pilot is a Level 4 autonomous driving extension of the Highway Chauffeur [11]. The driver is still in the driver's seat but the vehicle can perform highway driving completely autonomously without the need for driver interaction. Highway Pilots are expected to enter the market after 2022 [12].

In Level 5 driving automation, there is no need for a driver as the vehicle takes care of all driving tasks autonomously. It is clear that SAE Level 4 and Level 5 autonomous vehicles have to be capable of making their own decisions based on situational awareness using perception sensors and decision-making algorithms to satisfy the fixed mission of following the highway between initial and final destination locations. This includes planning their route once the destination point is specified and taking care of path planning, path-tracking control, and collision avoidance maneuvering, if needed, autonomously. This same approach is also needed for the lower speed autonomous driving in urban city environments which is a much more complicated situation due to the many other actors like vulnerable road users being present and more unexpected situations being likely to occur. This book treats path planning, path tracking control, and collision avoidance maneuvering for both urban and highway autonomous driving and also treats pedestrian collision avoidance of autonomous driving in the context of the urban application.

Automated driving shuttles in smart cities that are used for solving the first-mile and last-mile problem are other well-known, emerging examples of autonomous road vehicles [11]. These shuttles operate at relatively lower speeds which definitely improves safety levels while also creating a traffic bottleneck around them. In comparison to limited access highway operation, these shuttles operate in significantly less-structured environments with unpredictable interaction with vulnerable road users such as pedestrians, bicyclists, and scooters. The roads they use involve pedestrian crosswalks, intersections with or without traffic lights, roundabouts, and sharper turns as lower speed of operation is possible. Successful applications of these low-speed autonomous shuttles exist in fixed routes. The whole route needs to be mapped in advance and extra landmarks in the form of signage have to be added in some cases as scan matching of the recorded map is used for localization of these autonomous shuttles. Level 4 like autonomous driving of these shuttles is achieved during the segments of the route without intersections and unexpected interactions with other road users. The safety driver takes over control of the vehicle in intersections and during unexpected events. This is called assisted autonomy and is currently necessary for safe operation. True Level 4 autonomous driving capability of these low-speed urban environment autonomous vehicles is expected to be realized in the near future.

The most fundamental task of an autonomous vehicle is the ability to plan and follow a path while avoiding collisions. Path planning is optimized to make sure that the resulting trajectories have comfortable motion with limited acceleration and jerk. Uncertainties in environmental conditions, vehicle dynamics, vehicle load, and load distribution and the range of required speed from very low speeds for urban driving to highway driving speeds require the path tracking and collision mitigation controls to be robust. The motivation of this book is to contribute to this very important area of autonomous driving by presenting recent research results in path planning and robust path-tracking control. Robustness is achieved through two different approaches. The first one is regulation of the path following dynamic model to reject the uncertainties and disturbances and to handle the variable time delays that are present. The second approach is to use a robust feedforward and feedback controller combination to achieve guaranteed performance. The presence of static or moving obstacles such as other cars, pedestrians, and bicyclists is also treated by presenting methods for modifying the path to avoid such collisions in realistic applications. The methods presented in the book are applicable in real life, having been tested in a realistic hardware-in-the-loop simulation environment and in road testing with a research-level autonomous vehicle in addition to the usual model-in-the-loop simulations.

1.2 History of Automated Driving

Contrary to popular belief, the origins of autonomous driving and automated vehicles go back all the way to the 1920s. Radio-controlled cars were the novelty in the 1920s while 1960s and 1970s saw the emergence of cable-controlled cars, actually and unknowingly taking a step backwards. Computer-controlled cars resembling today's autonomous vehicles started emerging in a very rough form in the 1980s and 1990s. In the first driverless car experiments of the 1920s, an antenna was mounted on the car which was driven by an external operator using radio signals, much like radio-controlled toys. It should be noted that this remote operation forms the basis of some current driverless vehicles being followed by a second vehicle whose operators can take over control and intervene, if necessary. The presence of a safety operator, whether in a nearby other vehicle or in the driver's seat, is one of the current major limitations of autonomous driving. The car in the 1920s example obviously had to be in the field of vision of the external operator and also had to be operated at low speeds. A totally different approach was taken starting in the 1960s to get rid of the safety driver. Cables carrying electricity and generating a magnetic field were embedded in the roads. A downward-looking magnetic pickup in the car was used to follow this magnetic field much like the way toy cars follow reflective tape fixed on the ground. This approach obviously did not work in the long run due to the tremendous work and cost required for the necessary road infrastructure. The first computer-controlled cars that were developed later used camera and later also radar to do lane keeping and speed alterations much like today's lane-centering and ACC systems to automatically follow roads and cars. The computational power available was very limited as compared to what is available now with our robust operating systems and our fast CPUs and GPUs with large memory and storage capability.

The first documented radio controlled car was produced in 1925 by Houdina Radio Control, a radio equipment firm. The car called the Linrrican Wonder traveled through a traffic jam in New York City and was controlled by a transmitting antenna. The car behind this "phantom auto" sent signals to the antenna, and the signals worked on small electric motors that actuated the necessary pedals and steering. The Linrrican Wonder was obviously also a drive-by-wire vehicle. It was driverless in that the driver was in the following vehicle. The actuator positions of this following vehicle were mimicked by the driverless vehicle in a master–slave configuration. This is similar to the master–slave configuration that is used in some platooning applications with the difference that the master is the following vehicle and not the lead one which will obviously introduce unnecessary phase lag to the coupled system due to the communication time delay involved and this delay will destabilize the system at higher speeds. Nevertheless, this radio-controlled phantom lead car followed by the second control signal-generating car is an early example of automated platooning. The control signals rather than vehicle acceleration are sent to the next vehicle as is proposed in the recent connected cruise control (CC) concept. The human operator in the following car also acts like a cloud computer that computes and relays the necessary control commands to the driverless lead car. This radio-controlled driverless car is illustrated in Figure 1.2. Note that a similar procedure of having a second car with operators follow a truly driverless car with passengers was used recently for safety purposes [13]. This radio-controlled driverless vehicle concept from the 1920s also has similarities to the currently used phantom driver concept which is used to remotely operate a truly driverless vehicle in emergency situations that the autonomous driving system cannot handle [14]. The remote operator has full access to the sensor data and surround view from

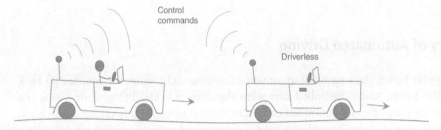

Figure 1.2 Illustration of radio-controlled driverless car of 1920s.

the driverless car and can operate the car using his/her steering and pedal inputs like manual driving. The variable communication delays involved create possible stability problems that need to be handled just like those that occur in telemanipulation. These delays are expected to become much smaller in magnitude with 6G communication which is the driving point behind current collaborative perception and awareness systems.

There have also been successful proof-of-concept type implementations of Intelligent Transportation Systems (ITSs) technology dating back to 1925. Charles Adler was named the man who invented Intelligent Traffic Control a century too early [15]. He embedded magnetic plates in the road at a point before the road led into a sharp curve. He also prepared a car with a speed governor that would be activated as the car drove over the magnetic plate such that the vehicle engine would be commanded to slow down to 24 km/h. This is a very early hardware implementation of Curve Speed Warning [16] and automatic speed reduction system where the whole system has been implemented as a hardware solution.

The General Motors *Futurama* exhibit at the 1939 World's Fair in New York is viewed by many as the first large-scale ITS and highway automation concept and vision [17]. The vision of *Superhighways* was demonstrated in this exhibit where the visitors would sit and watch a scaled down replica of a future city complete with highways and guided cars. This was a hard 3D model version of the current game engine rendered environments with traffic cosimulation [3, 18–21]. These guided cars were navigated automatically using radio control and could enter curves at speeds up to 50 mph (80 km/h). This vision of Superhighways was presented as what would happen in 1960. In compliance with this idea, there was a desire in the automotive industry to have cable buried in lane centerlines in these envisioned superhighways to create a magnetic field that could be followed automatically by self-driving vehicles. This approach was not adopted because of higher initial cost and higher maintenance cost during repairs. This vision was one of self-driving based on infrastructure which is exactly why it failed.

The current infrastructure-based technology that can be used for following lane centerlines is the use of a front-looking camera to detect and track the lanes in a lane-keeping assistance system. The problem with this technology is that the lanes cannot be detected when weather conditions degrade the camera performance or when the road is covered with snow, making the lanes invisible. Since the interstate roads were built without voltage carrying cable buried inside, automotive OEMs lost interest in self-driving as the required infrastructure was not there. Researchers and some states like California continued on with this Superhighway idea, which changed its name to Automated Highway Systems, until the 1990s, but while there were a lot of successful demonstrations and a large literature of research results, larger-scale use of this technology never happened as automotive companies did not want to develop cars dependent on nonexisting and costly infrastructure and as there was also no demand from car buyers. Nevertheless, different ways of embedding passive signals to be followed inside roads continued until recent years in the form of equally spaced metallic pins, reflective tape, and more recently smart paint [22, 23]. Out of these, smart paint uses nanoparticles embedded inside normal paint used in lane markings and can be detected with a simple sensor. It can also be detected if the road is covered with snow and offers a relatively cheap and highly accurate way of localization in campus like environments or geofenced urban areas for low-speed autonomous driving applications as the preview distance is constrained to be small.

According to reference [24], the first self-driving car was built and tested by the Radio Corporation of America (RCA) in 1957 on a 400 ft public highway in Lincoln, Nebraska. The steering wheel and pedals in this vehicle were replaced with a small joystick and an emergency brake while the vehicle speed and distance with the car in front were displayed on the dashboard. Self-driving was activated by pressing the Electronic Drive button and the car would follow its lane and adjust its speed automatically. RCA was working on this technology around the 1950s. During the same time frame around 1956, GM shared its vision of a futuristic car driving autonomously in an automated highway using a professional video in the Motorama auto show. Automated highway driving involved a control center directing traffic much like today's aircraft traffic control centers.

According to Wetmore [25] GM and RCA developed a scaled version of an automated highway system by 1953. They used this scaled automated highway system to investigate how electronics can be used for path-tracking and car following. Note that this is very similar to the scaled drive-by-wire vehicles with sensors like the F1TENTH vehicles [26, 27] that a lot of academic research groups use for in-the-lab studies of autonomous driving and decision-making. In 1958, GM built a prototype self-driving vehicle with a pickup in front that would sense the alternating current of a wire embedded in the road [25]. This was used for localization with respect to the path to be followed and for generating the required corrective steering action. The research vehicle could take turns automatically without driver steering intervention. By 1960, GM had tested its research vehicles in test tracks with automatic path-tracking, lane-change maneuvering, and car following. Since the technology used was based on cable and signals buried under the road, these research efforts stopped after the interstates were built without this infrastructure required for automated highways.

The first university based self-driving studies started around the 1960s just as the industry studies were ending. The same technology of a live wire either buried in the road or fixed on top of the road was used to steer a vehicle in a closed test area as shown in Figure 1.3 by Ohio State University researchers [28]. An analog computer was used to implement steering control and speed control as shown in Figure 1.4. This vehicle was converted into a drive-by-wire one and the analog joystick shown in Figure 1.5 was developed and used in the car for operator intervention. Turning the handle of the joystick left or right was used for steering in the same direction. Moving the joystick handle forward applied throttle while moving it backwards applied braking. In later years, longitudinal car following was implemented by actually connecting the two cars by wire and communicating the necessary information like distance and speed difference through this hard wire. Connected vehicles (CVs) actually meant being physically connected during 1960s. As compared to today's wireless networked car following systems, this wired and truly connected version obviously did not have the wireless communication delays that limit the minimum achievable time gaps of present platooning systems and cooperative ACC.

While the historical self-driving car work presented until this point focused on automating the longitudinal and lateral motions of the on-road vehicle, its controls, some very basic radar perception and road-fixed cable based localization; work on autonomy or intelligence, i.e. decision-making, was not being pursued owing mostly to the structured nature of highway driving. On the other hand, the mobile robotics community was using low-speed mobile robots with simple mechanical designs to use camera sensors for perception of the environment and

Figure 1.3 First university based self-driving study by Ohio State University researchers. Source: Fenton and Olson [28]/with permission from IEEE.

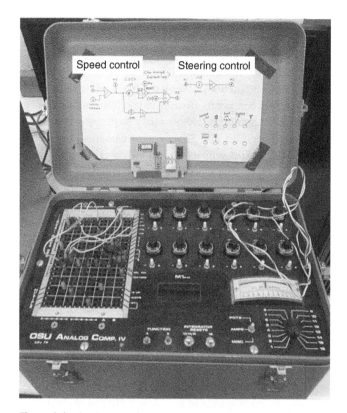

Figure 1.4 Analog steering and speed control of 1960s Ohio State University self-driving vehicle. Source: The Ohio State University.

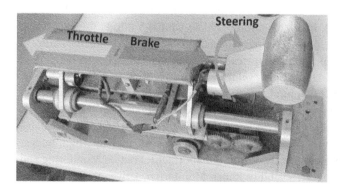

Figure 1.5 Joystick for drive-by-wire interface of Ohio State University self-driving vehicle from the 1960s. Source: The Ohio State University.

scene understanding to be used in decision-making for autonomously navigating through previously unknown obstacles by planning and executing a collision free path. Unlike the highway driving application, these mobile robot studies were geared toward rovers to operate in the moon and much later in Mars and, thus, required a high level of autonomy in a highly unstructured and uncertain off-road environment. The requirement of remote control of these mobile robots from earth also introduced considerable time delay on the order of 2.5 seconds for the shortest tele-distance example into their feedback control loop. The Stanford cart was the first prototype of these mobile robots introduced in 1961 [29]. Much work was done using this research platform and its extensions in the 1970s. The Stanford Cart used cameras to detect and follow a solid white line on the ground which is similar to the vehicles following a live cable buried inside the road.

The Stanford Cart was followed by the Artificial Intelligence Center of Stanford Research Institute's (SRI) mobile robot called Shakey. The Cart and Shakey were the first mobile robots and also the first self-driving vehicles that used digital computers instead of electronics and analog computers. These computers were like the cloud computers of today as they were using very large mainframe computers through radio links. Both the Cart and Shakey used cameras to perceive the environment as compared to magnetic field pickup sensors and radar being used by their on-road vehicle counterparts of the same timeframe. The Cart was able to follow white lines and Shakey used image processing to detect large prismatic objects. The mechanical structure of Shakey looked a lot like the mobile robots of today. Current mobile robots, of course, have very high-embedded computational power and use three-dimensional lidar along with cameras. They also have relatively high accuracy global positioning system (GPS) and inertial measurement unit (IMU). The image processing and decision-making algorithm development approach of the mobile robotics approach and the highway ready vehicle platforms, motion controls, and radar sensors of the on-road self-driving approach united in an evolutionary manner in the decades to follow the 1970s to form the basis of current autonomous driving vehicles with the peak results being achieved in the DARPA Challenges of 2000s.

A rough combination of both approaches resulted in the autonomous vehicle of Japan's Tsukuba Mechanical Engineering Laboratory. This vehicle that looked like a large multiwheeled mobile robot with an armored-vehicle like body shell first used two cameras for perception of white lines on the road. Analog computers were used for signal processing as digital computers were still too large to fit inside the vehicle. The car was able to reach speeds of to 30 km/h (18.6 mph) which is close to the 25 mph speed limit of current autonomous shuttles used in urban areas. The digital computer problem was solved in the late 1980s as they were able to fit within a van. In Germany, in the beginning of the 1980s, a Mercedes-Benz van with cameras and other sensors used a digital computer for self-driving control [30]. In 1986, another vehicle named VaMoRs from the same group drove by itself with speeds up to 96 km/h. being achieved in 1987. It is clear that the ability to use digital computers and use of in-vehicle sensors instead of infrastructure-based sensing sparked the interest of automotive companies in self-driving vehicles again.

During 1987 to 1995, the European Union funded the Prometheus (PROgraMme for a European Traffic of Highest Efficiency and Unprecedented Safety) project led by major European automotive OEMs and research organizations [31]. A 1758 km self-driving trip from Munich, Germany, to Copenhagen, Denmark, and back was the end result of this large scale and very high budget project. In 1994, the project demonstrated in Paris: Vision Enhancement, Friction Monitoring and Vehicle Dynamics, Lane Keeping Support, Visibility Range Monitoring, Driver Status Monitoring, Collision Avoidance, Cooperative Driving, Autonomous Intelligent Cruise Control, Automatic Emergency Call, Fleet Management, Dual Mode Route Guidance, Travel and Traffic Information Systems. These are all parts of a Highly Automated Driving (HAD) system. The Paris demo in 1994 also involved the twin automated vehicles VaMP and VITA-2 self-driving for more than 1000 km on a multiple lane highway in traffic at speeds up to 130 km/h. Self-driving included lane keeping, car following of the two automated vehicles, and lane changes for overtake maneuvers.

The situation in the United States was much different than the one in Europe as automotive OEMs had lost interest in self-driving technology. During 1980s, the research work at the Ohio State University ended due to lack of funding. Starting around 1986 and continuing until present, the California Partners for Advanced Transit and Highways (PATH) carried out research on automated highway systems first and on ITS in later years. PATH is a collaboration between the California Department of Transportation (Caltrans), UC Berkeley, other public and private academic institutions and private industry. While the PATH program was well funded and was very successful in the deployment of intelligent vehicle technologies such as convoy driving, merging, leaving convoys, it suffered from not being backed up by automotive OEMs. As a result, some early ideas failed and the research results were not adopted by the automotive industry until much later. In 1995, two researchers from Carnegie Mellon University's Robotics Institute used a minivan they called NavLab 5 to drive autonomously in a 2850 mi trip from Pittsburgh to San Diego [32, 33]. Their self-driving system called Rapidly Adapting Lateral Position Handler (RALPH) was a camera-based system and did 98% of the driving. The technology had been developed in CMU's Robotics Institute since 1986. NavLab 5 used a desktop-like computer, a camera, a GPS sensor and a high accuracy fiber-optic gyro sensor. NavLab 5 had lane keeping, automatic lane changing, blind spot monitoring and decision making capabilities. Two years later, the demonstration called Demo'97 was organized by the National Highway Systems Consortium (NHSC) in 1997 with participants invited from research organizations, universities, and automotive manufacturers and suppliers [34]. Separate, special purpose lanes in a highway in California were closed for the demo which mainly consisted of platooning and related technologies to demonstrate the efficacy of Automated Highway Systems. Within the same timeframe in 1994, a computer vision lab called VisLab in Italy developed a self-driving vehicle they called ARGO [35]. The camera-based self-driving capability of ARGO was tested for more that 2000 km on Italian highways with regular traffic, and ARGO was in self-driving mode 94% of the time during this deployment.

The 2000s were a revolutionary timeframe for autonomous vehicles with the DARPA autonomous driving challenges taking place. The DARPA Grand Challenges took place in the Mojave Desert in the United States during 2004 and 2005 [36, 37]. The route was 241 km alongside Interstate 15. The first DARPA Grand Challenge in 2004 was not successful as none of the autonomous vehicles were able to finish it. The longest distance traveled was 11.78 km by the Red Team from CMU Robotics Institute [38]. There was a qualifying round at the California Speedway test course at Fontana, California, where the teams showed their vehicle and approach to the judges to be selected for the actual event. The actual route, out of many possible choices and GPS waypoints to be followed, was given to the teams shortly before the challenge so that they could not create a route specific solution in advance. The waypoints that were given to the teams were so dense that a path planning algorithm was not even needed, considering also the necessity of using lower speeds because of the off-road like terrain. Fifteen vehicles were allowed to take part in the 2004 challenge but two of them had to be withdrawn before the event. The challenge was scheduled to run for ten hours but only four of the participant vehicles were operational after the first three hours. Most vehicles had mechanical failures or GPS problems. Some of them were stopped by the judges as their algorithms got stuck. The Red Team vehicle that went furthest also went off course at the end and got stuck in an embankment. While the 2004 DARPA Grand Challenge was not successful operationally, there were a lot of lessons learned and a lot of collected data to work with.

When the challenge was repeated in 2005, the new off road course was 212 km long [39]. Except for one team, all of the 23 finalists were able to go beyond the 11.78 km maximum distance traveled in the 2004 event autonomously. Four of the autonomous vehicles completed the course on time while a fifth one, a large truck, finished over the ten-hour limit. The 2005 route had three narrow tunnels and 100 high curvature turns as compared to the wider and less curved route of the 2004 challenge. Stanley from Stanford University was the winner of the 2005 DARPA Grand Challenge [40]. The Grand Challenges were followed in 2007 by the DARPA Urban Challenge that took place in an Air Force base in California [41]. The course was 96 km long, representative of an urban driving environment and had to be completed within six hours. There were manually driven vehicles with professional drivers who circulated parts of the route to re-create the urban driving environment. DARPA organized the teams into two tracks and provided $1 M funding to teams in Track A most of which teamed up with automotive OEMs and technology companies. Not all of the teams were invited to the finals based on their semifinal performance. Tartan Racing's (CMU) Boss finished in first place, followed by Stanford Racing's Junior [42, 43]. Only six teams were able to finish the whole course and out of those only four teams were able to finish in time. The average speed of the first three teams was about 13–14 mph which is similar to the speeds of current autonomous shuttles operating in urban environments. The Massachusetts Institute of Technology (MIT) and Cornell teams had their famous autonomous vehicle (AV) accident in this event, but they were both able to finish the challenge, nevertheless [44]. In the DARPA Grand Challenges, all teams had similar sensor, computational system and software configurations with the gap, increasing in favor of the winning teams in the 2005 challenge. The difference in sensor, computational platform and AV software between the winning teams and the others showed a major gap in the Urban Challenge where some of the winning teams had developed perception, detection/tracking, planning, path-tracking, and visualization routines similar to the ones used today which gave them a significant advantage.

The 2007 DARPA Urban Challenge and the subsequent interest and work of Google using the developed technology for recording images and building maps more efficiently have laid the framework for today's autonomous driving technology. The low-speed, self-driving vehicles that Google developed and used for map-building purposes had a very strong impact on the public image of AV technology. Even though automotive OEMs, suppliers, research centers, and universities had already significantly developed autonomous driving technology, vehicle owners had not shown a big enthusiasm for and interest in technology. Very interestingly, the successful implementation of existing AV driving methods by a search engine company was more impactful in making people believe that AV technology is for real and works. With public acceptance, belief, and expectation, autonomous driving has since seen extensive research and development investments.

Successful deployments of AV technology continued after the DARPA Urban Challenge of 2007. A significant one of these deployments was the VisLab Intercontinental Autonomous Challenge funded by the European Union. A convoy of vehicles, two of them being AVs that follow each other, started from Parma, Italy in 2010 and drove semiautonomously all the way to Shanghai, China [45]. The first vehicle in the convoy was driving semiautonomously while the second vehicle followed it autonomously. They were followed by other vehicles used for support and maintenance. The 2013 autonomous drive on the Bertha Benz Memorial Route was an example of another successful autonomous driving deployment after the DARPA challenges. The course taken by the autonomous vehicle was 103 km long that included both rural roads and major cities with a large variety of difficult traffic scenarios including intersections with and without traffic lights, roundabouts, and narrow passages with oncoming traffic [46].

On the cooperative autonomous driving side, the Grand Cooperative Driving Challenge (GCDC) took place in 2011 in Helmond, Netherlands, on a closed highway segment where eight different teams ran multiple heats of cooperative driving with autonomy in the longitudinal direction [47]. While cooperative driving deployments had taken place before, this was the first one where the two side-by-side platoons had vehicles from different teams that used different vehicles ranging from a Smart to trucks. The convoys had different vehicle pairings in each heat and drove autonomously (longitudinally) and cooperatively while performing convoy merging, handling traffic lights and responding to speed limits autonomously, using IEEE 802.11p DSRC V2X communication. As the number of autonomous vehicles on public roads increases, they will have to communicate with each other, with the infrastructure and driving in a coordinated fashion. The GCDC 2011 was repeated in 2016 with both longitudinal and lateral autonomy. After only a decade since the GCDC 2011, DSRC technology for V2X communication is being replaced by cellular communication. Truck platooning and automated-driving trucks are applications of this technology that are currently being used on public roads.

More recent deployments of AV and Connected Vehicles (CVs) technology are too many to report. Of interest is the Roborace since it is a truly driverless race in which all teams use the same vehicle and hardware and software platforms, making it a race between different AV driving algorithms. The 2016 Smart City Challenge in the United States is also an interesting recent example of many AV/CV deployments that recently ended in Columbus, Ohio, as the Smart Columbus project [3]. AV shuttles were successfully operated in two different urban environments in Smart Columbus and more than 100 traffic lights and about 1500 vehicles were fitted with road-side units (RSUs) and on-board units (OBUs), respectively, with collected data being available on a cloud server.

1.3 ADAS to Autonomous Driving

The continuity of developments in active safety systems and ADAS in the automotive area have resulted in the basis of current HAD and Automated Driving technologies that are gradually being introduced to series-produced road vehicles. The earliest active safety system is anti-lock brakes (ABS) that prevent locking of wheels during sudden braking especially on slippery surfaces while also aiming to keep steerability. This is a very important feature for obstacle avoidance systems in path-tracking control of autonomous vehicles. Traction control systems (TCSs) work in the opposite direction of wheel slip and aim at avoiding spinning of the wheels during sudden applications of driving torque, again especially on slippery road conditions. ABS focuses on all wheels of the vehicle by individually braking them for optimizing slip while TCS applies individual wheel braking to just the drive wheels along with drive torque reduction. Electronic stability control uses the same hardware to reduce drive torque and brake wheels individually to reduce speed and apply steer-by-braking to regulate the yaw rate of the vehicle to correspond to the driver applied steering. ABS/TCS/ESC is currently available as a bundle in current cars and ABS and ESC are mandatory. While TCS is not mandatory, it is always present since it is very easy to implement once ABS and ESC are present and since customers demand its presence. While braking-based ESC was explained here, steering-based ESC is also possible with a combination of individual wheel braking and steering giving best results [48].

Years of development of ABS/TCS/ESC systems have resulted in the possibility of easily implementing automated braking and throttle systems in current vehicles for longitudinal automation of motion. Wheel speed sensors used in ABS/TCS are available to the autonomous driving system also. The accelerator pedal position sensor(s) and the brake switch sensor or the brake pedal position sensor in a hybrid or fully electric vehicle are also available to the autonomous driving system and signal driver takeover request in the longitudinal direction. In addition, the ESC system introduces a steering wheel position and rate sensor, a yaw rate sensor and an accelerometer with longitudinal and lateral acceleration all of which are accessible using the vehicle controller area network (CAN) bus to the autonomous driving controller which is also a vehicle dynamics controller but with perception, situational awareness, and decision-making. It is expected that future autonomous vehicle path-tracking controllers will take over the tasks of ABS/TCS/ESC subsystems. The current technology uses a handover of control to ABS/TCS/ESC systems when they are activated.

These functions should be an integral part of the AV trajectory tracking controller with seamless operation.

Power steering systems and electric or electric-assisted power steering systems are now common in road vehicles as they are also necessary for automatic parking and lane-keeping assistance ADAS systems. While power-assisted steering requires the electric motor to operate in torque control mode, these electric motors can also switch to position control mode which is used for lane keeping and parking assistance. Using the appropriate CAN bus commands, the power steering actuator can be used to automatically steer the vehicle. For this purpose, the power steering actuation system should be designed to be able to also handle emergency lane change type maneuvers that may be required in autonomous driving. Autonomous driving will also require an automatic transmission with the ability to change the transmission state electronically through CAN bus commands. This also includes transmission gear changes for optimum performance. In most research AVs and retrofitted vehicles for autonomous operation, the vehicle's inherent automatic transmission controller takes care of the gear switching and the AV driving system can only switch state between neutral, drive, park, and reverse.

The basic Cruise Control (CC) system has been available for a long time and has advanced from a simple mechanical type control to an advanced speed controller for the vehicle. Most current vehicles are easily equipped with ACC which uses longitudinal automation and a front-looking camera/radar combination to follow a slower target vehicle in front at the desired time gap selected by the driver. These systems are now capable of operating all the way down to a complete stop (although requiring a minimum speed to be initiated first). Along with a Lane Keeping System (LKS), an ACC system provides a very basic automated driving capability in highways. As these two ADAS systems are very well calibrated by automotive OEMs, they work very well in driving situations with the decision-making still carried out by the driver. Improving this system into a traffic jam assistance which also works at very low speeds down to zero result in a very basic highway driving automation and forms the basis of some recently announced L3 driving capable series produced vehicles [1]. The natural evolution of automated driving from ADAS systems is a much safer approach as compared to retrofitting a vehicle for automated operation and re-designing the control system since automotive OEMs and suppliers have already devoted extensive development and testing effort and experience to these systems which are also highway legal. Current autonomous shuttles, for example, cannot handle speeds above 25 mph on exurban roads with 35 or 45 mph speed limits, also due to the fact that it is harder for them to map those parts of roads due to the lack of enough landmarks and due to memory limitations. ADAS systems that already exist in series produced vehicles can easily take care of automated driving in those segments of the route without the need for scan matching of a high-definition 3D lidar map.

Automatic Emergency Braking (AEB) is another ADAS and safety feature that is easily available in current series produced vehicles. It requires short-range radars also to determine dangerous oncoming target vehicles at close range. Some implementations of AEB can even steer the vehicle slightly if a collision cannot be avoided by braking alone. If the AEB system had not been disengaged in the 2018 Uber accident, for example the accident could have been avoided or could have been nonfatal [49]. Lane Change Assistance Systems (LCASs) use short range side radars to determine a collision risk during lane changing. It is clear that the presently available ADAS systems like LKS (also denoted LKAS for Lane Keeping Assistance System), ACC, AEB, LCAS, and their extensions can form the basis of an automated driving system. All the actuators, sensors, and basic lower-level control systems are already available and need to be complemented by or upgraded to a more powerful computing system with integrated, more advanced algorithms for situational awareness and decision-making. Navigation systems are also very common and easy to add to series production vehicles. An electronic horizon system can also be used instead of the navigation system. The navigation system or electronic horizon can be used to determine the route or path once the destination is selected.

1.4 Autonomous Driving Architectures

An autonomous driving vehicle architecture consists of the actuation, low-level controls, perception and communication sensors, localization sensors, localization, sensor fusion and situational awareness, computational system, and higher-level decision-making layers. One can also take a look at the hardware and software architectures separately. The software architecture will then comprise of the AV driving functions. These driving functions have to include localization to answer: "Where am I?," user input to answer: "What is my destination," path planning to answer: "How do I get there" and collision avoidance maneuvering to answer: "How do I avoid obstacles along my route?" Some basic architectures and rule-based approaches to decision-making are presented in Chapter 3 of this book. The rule-based approach for decision-making uses Finite State Machines (FSMs). An alternative approach is to use data-driven approaches like deep learning or reinforcement learning [50]. While the literature has an abundance of papers on data driven approaches, they unfortunately neglect the dynamic nature of the vehicle and its limitations, resulting in suboptimal path following performance accompanied by discontinuous and jumpy steering and nonsmooth motion especially when end-to-end data-driven approaches are used. The recommended alternative is to use an end-to-mid approach where the path-tracking control is taken outside the data-driven part of the implementation and, hence, made vehicle platform agnostic [51]. This book fills an important gap in this area by introducing trajectory planning and tracking methods that respect the dynamic nature and limitations of an actual road vehicle and result in physically implementable systems with smooth AV motion characteristics.

1.5 Cybersecurity Considerations

The literature review on path modeling, planning, and path-tracking control is presented in detail in the relevant chapter. We will talk briefly on cybersecurity concerns and issues. Autonomous vehicles can be hacked just like computers, and cybersecurity issues should not be underestimated. V2X connectivity that is present in AVs is an important source of cybersecurity concerns. The Security Credential Management System (SCMS) which is a security certificate is used in connected AVs to overcome cybersecurity issues due to V2X connectivity. The connected AV will trust nearby communicating agents that have the correct security certificate which like all certificates can be faked. It is, thus, possible to send wrong MAP and SpaT messages from a connected intersection resulting in the AV making wrong decisions on the location and geometry of the intersection and the signal phase. A very simple example is fooling an AV into a green phase even though the actual phase is red. It is also possible to introduce fake or ghost vehicles or sudden obstacles by sending fake locations in the Basic Safety Message (BSM) in V2V communication. This may result in the vehicle performing an emergency stop or emergency evasive maneuver even though there is no need for it. It is evident that such problems cannot simply be resolved with certificates and encryption alone. The use of a supervisory copilot that continuously monitors the connected AV and does a safety check of its planned motion is what is proposed here as a solution. This supervisory copilot (not a person but another autonomous system) continuously checks the perception sensors and compares with the communicating agent data to determine whether the communicated information is trustworthy or not. Once the immediate future part of the path including modifications for obstacle avoidance are computed, they can be checked using control barrier functions, for instance, for feasibility and also to avoid cybersecurity issues due to malicious information [52]. Control barrier certificates can also be used for the same purpose [53].

The same approach has to be used for the perception sensors whose data is fused. Support Vector Machine (SVM) computations can be used for this purpose also [54]. Note that some perception sensors can be fooled just like they are artificially generated in hardware-in-the-loop (HIL) simulations. Fake Doppler signals or lidar signals can be physically sent to misguide radar or lidar sensors. Light sources can be used to blind out camera sensors. GPS signals can be jammed or spoofed. The AV has to rely on redundant sensor data to identify the presence of an issue and in extreme cases, report the situation, pullover and stop autonomous operation.

Even though several articles have reported hacking of the vehicle CAN bus in the past, this was only possible as an interface device was physically attached to the CAN bus connector which is highly unlikely in practice [55]. The infotainment system in newer cars communicates with the outside worlds and is, thus, open to attacks. Indeed, most of the car hacking examples reported until now have been achieved through vulnerabilities in the infotainment system. If the infotainment system has the capability of access to the vehicle CAN bus, it is possible to take over control of the vehicle actuators and functions and command them remotely. This is a very dangerous situation. The main dangerous outcome of all malicious attacks on a road vehicle, especially a connected and autonomous one, is the modification of the path of the vehicle and the associated controls which may lead to fatal accidents. This means that the path-planning system and path-tracking controls should continuously check the planned path and the actual path tracking against the vehicle mission of reaching the final destination and for obstacle avoidance and for obeying rules of traffic.

1.6 Organization and Scope of the Book

This book is organized in to eight chapters. The second chapter following this introductory chapter is on the tire and vehicle dynamics, path and path-tracking modeling. The third chapter is an overview of simulation, experimentation, and parameter estimation. The fourth chapter treats mathematical modeling of the path including the generation of such models using data in the form of waypoints to be followed. Chapter 5 is on collision free-path planning and introduces methods of modifying the path to avoid obstacles. Chapter 6 introduces model regulation or the disturbance observer loop to reject road curvature as a disturbance in path-tracking control. Chapter 7 uses robust gain scheduled parameter space design and linear matrix inequality (LMI) control for path tracking. The book ends in Chapter 8 with conclusions.

1.7 Chapter Summary and Concluding Remarks

This is the first chapter of the book and forms an introduction to autonomous vehicle path planning and path-tracking control. The chapter started with an introduction to autonomous vehicles and the motivation for the book as collision free-path planning and path tracking are the most essential functions of autonomous driving. After the introduction and motivation, a brief history of autonomous driving is presented. It is seen that the two separate research directions of road vehicle automation and autonomy of mobile robots have merged together to form the basis of modern-day autonomous vehicles. Active safety systems and ADAS for series produced road vehicles are discussed briefly to show the safe and gradual approach to autonomy and intelligence taken by the automotive industry. Brief information on autonomous driving architectures and cybersecurity issues are the last topics treated in this chapter.

References

1 Anonymous, "New Level 3 Autonomous Vehicles Hitting the Road in 2020," 2020. [online]. Available: https://innovationatwork.ieee.org/new-level-3-autonomous-vehicles-hitting-the-road-in-2020/.

2 M. M. Nesheli, L. Li, M. Palm, A. Shalaby, Driverless shuttle pilots: lessons for automated transit technology deployment, *Case Studies on Transport Policy*, vol. 9, no. 2, 2021, pp. 723–742, ISSN 2213-624X, https://doi.org/10.1016/j.cstp.2021.03.010.

3 L. Guvenc, B. Aksun-Guvenc, X. Li, A. Doss, K. Meneses-Cime, and S. Gelbal, "Simulation Environment for Safety Assessment of CEAV Deployment in Linden," arXiv preprint arXiv:2012.10498, 2020.

4 X. Li, S. Zhu, S. Gelbal, M. Cantas, B. Guvenc, and L. Guvenc, "A unified, scalable and replicable approach to development, implementation and HIL evaluation of autonomous shuttles for use in a smart city," in SAE World Congress and Experience, Detroit, 2019.

5 B. Templeton, "AutoX opens real robotaxi service in China to the general public," Forbes, 2021.

6 Anonymous, "Roborace," 2021. [online]. Available: https://roborace.com/.

7 E. Ackerman, "Nuro raises $92 million for adorable autonomous delivery vehicles. IEEE Spectrum, 30 January 2018.

8 R. Bishop, "U.S. States are allowing automated follower truck platooning while the swedes may lead in Europe," Forbes, 2020.

9 L. Guvenc, B. Aksun Guvenc, and M. Emirler, "Connected and autonomous vehicles," in H. Geng *Internet of Things and Data Analytics Handbook*, 2017, pp. 581–595: Wiley, New York.

10 S. Luca, A. Serio, G. Paolo, M. Giovanna, P. Marco, and C. Bresciani, "Highway Chauffeur: state of the art and future evaluations: implementation scenarios and impact assessment," in International Conference of Electrical and Electronic Technologies for Automotive, Milan, Italy, 2018.

11 S. Gelbal, N. Chandramouli, H. Wang, B. Aksun-Guvenc, and L. Guvenc, "A unified architecture for scalable and replicable autonomous shuttles in a smart city," in IEEE International Conference on Systems, Man, and Cybernetics, Banff, Canada, 2017.

12 R. Baldwin "2022 Volvo XC90 Will Get Autonomous Highway Pilot Hardware with Lidar," Car and Driver, 2020.

13 A. J. Hawkins, "Waymo's driverless car: ghost-riding in the back seat of a robot taxi," The Verge, 2019.

14 A. Davies, "Self-driving cars have a secret weapon: remote control," Wired, 2018.

15 L. Vinsel, "The man who invented intelligent traffic control a century too early," IEEE Spectrum, 21 July 2016.

16 Anonymous, "Curve Speed Warning," ITERIS, [Online]. Available: https://local.iteris.com/cvria/html/applications/app13.html#tab-3 [Accessed 20 May 2021].

17 E. Kalan, "The original Futurama: the legacy of the 1939 world's fair," Popular Mechanics, 11 March 2010.

18 L. Guvenc and M. R. Cantas, "Customized co-simulation environment for autonomous driving algorithm development and evaluation," in SAE World Congress and Experience, Detroit, 2021.

19 X. Li, S. Zhu, B. Aksun-Guvenc, and L. Guvenc, "Development and evaluation of path and speed profile planning and tracking control for an autonomous shuttle using a realistic, virtual simulation environment," *Journal of Intelligent and Robotic Systems*, vol. 101, no. 2, pp. 1–23, 2021.

20 K. Meneses-Cime, G. Dowd, L. Guvenc, B. Aksun-Guvenc, A. Mittal, A. Joshi, and J. Fishelson, "Hardware-in-the-loop, traffic-in-the-loop and software-in-the-loop autonomous vehicle simulation for mobility studies," in SAE World Congress and Experience, Detroit, 2020.

21 X. Li, A. Doss, B. Guvenc, and L. Guvenc, "Pre-deployment testing of low speed, urban road autonomous driving in a simulated environment," *SAE International Journal of Advances and Current Practices in Mobility*, vol. 2, pp. 3301–3311, 2020.

22 C. Chan, B. Bougler, D. Nelson, P. Kretz, H. Tan, and W. Zhang, "Characterization of magnetic tape and magnetic markers as a position sensing system for vehicle guidance and control," in American Control Conference, Chicago, IL, 2000.

23 M. Wood, "How a Paint Developed at Ohio State Can Make Cities Safer and Smarter," BTN, 2018. [Online]. Available: https://btn.com/2018/04/20/how-a-paint-developed-at-ohio-state-can-make-cities-safer-and-smarter-btn-livebig/.

24 E. Ackerman, "Self-Driving Cars Were Just Around the Corner—in 1960," IEEE Spectrum, 31 August 2016.

25 J. M. Wetmore, "Driving the dream – the history and motivations behind 60 years of automated highway systems in America," *Automotive History Review*, vol. 40, pp. 4–19, 2003.

26 V. S. Babu and M. Behl, "f1tenth.dev – an Open-source ROS based F1/10 autonomous racing simulator," in IEEE 16th International Conference on Automation Science and Engineering, Hong Kong (virtual conference), 2020.

27 Anonymous, "F1TENTH,"[Online]. Available: https://f1tenth.org/.

28 R. Fenton and K. Olson, "The electronic highway," IEEE Spectrum, pp. 60–66, July 1969.

29 H. P. Moravec, "The stanford cart and the CMU rover," *Proceedings of the IEEE*, vol. 71, no. 7, pp. 872–884, 1983.

30 J. Delcher, "The man who invented the self-driving car (in 1986)," Politico, 19 July 2018.

31 M. Williams, "The PROMETHEUS programme," in IEE Colloquium on Towards Safer Road Transport - Engineering Solutions, London, 1992.

32 J. Togyer, "Then and Now: The 2,850-mile, no-hands road trip," CMU Computer Science Department, 28 October 2015. [Online]. Available: https://csd.cmu.edu/news/then-and-now-2850-mile-no-hands-road-trip.

33 P. Batavia, D. Pomerleau, and C. Thorpe, "Overtaking vehicle detection using implicit optical flow," in Proceedings of Conference on Intelligent Transportation Systems, Boston, MA, 1997.

34 J. Hedrick, "Nonlinear controller design for automated vehicle applications," in UKACC International Conference on Control, Swansea, UK, 1998.

35 A. Broggi, "History of AHS in Italy and future issues," in IEEE International Conference on Vehicular Electronics and Safety, Columbus, OH, 2008.

36 Anonymous, "The Grand Challenge," DARPA, [Online]. Available: https://www.darpa.mil/about-us/timeline/-grand-challenge-for-autonomous-vehicles.

37 Anonymous, "The DARPA Grand Challenge: Ten Years Later," [Online]. Available: https://www.darpa.mil/news-events/2014-03-13.

38 G. Seetharaman, A. Lakhotia, and E. Blasch, "Unmanned vehicles come of age: The DARPA grand challenge," *Computer*, vol. 39, no. 12, pp. 26–29, 2006.

39 Anonymous, "DARPA Prize Authority Report to Congress," DARPA, 2006.

40 S. Thrun, "Stanley: the robot that won the DARPA grand challenge," The 2005 DARPA Grand Challenge, Berlin, Springer, 2007.

41 S. Singh, M. Buehler, and K. Iagnemma, The 2005 DARPA Grand Challenge, Springer, 2007.

42 Anonymous, "DARPA Urban Challenge," DARPA, [Online]. Available: https://www.darpa.mil/about-us/timeline/darpa-urban-challenge.

43 S. Singh, M. Buehler, and K. Iagnemma, The DARPA Urban Challenge Autonomous Vehicles in City Traffic, Springer, 2009.

44 L. Fletcher, "The MIT – Cornell collision and why it happened," in M. Buehler, K. Iagnemma, S. Singh *The DARPA Urban Challenge*, Springer, pp. 509–548, 2009.

45 A. Broggi, P. Medici, P. Zani, A. Coati, and M. Panciroli, "Autonomous vehicles control in the VisLab intercontinental autonomous challenge," *Annual Reviews in Control*, vol. 36, no. 1, pp. 161–171, 2012.

46 J. Ziegler et al., "Making Bertha drive—an autonomous journey on a historic route," *IEEE Intelligent Transportation Systems Magazine*, vol. 6, no. 2, pp. 8–20, 2014, https://doi.org/10.1109/MITS.2014.2306552.

47 L. Guvenc, "Cooperative adaptive cruise control implementation of team mekar at the grand cooperative driving challenge," *IEEE Transactions on Intelligent Transportation Systems*, vol. 133, no. 3, pp. 1062–1074, 2012.

48 B. Aksun-Guvenc, T. Acarman, and L. Guvenc, "Coordination of steering and individual wheel braking actuated vehicle yaw stability control," in IEEE IV2003 Intelligent Vehicles Symposium Proceedings, Columbus, 2003.

49 M. Harris, "NTSB investigation into deadly uber self-driving car crash reveals lax attitude toward safety," IEEE Spectrum, 7 July 2019.

50 P. M. Kebria, A. Khosravi, S. Salaken, and S. Nahavandi, "Deep imitation learning for autonomous vehicles based on convolutional neural networks," *IEEE/CAA Journal of Automatica Sinica*, vol. 7, no. 1, pp. 82–95, 2020.

51 A. Doss and L. Guvenc, "Predicting desired temporal waypoints from camera and route planner images using end-to-mid imitation learning," in SAE World Congress and Experience, Detroit, 2021.

52 Y. Huang, S. Yong, and Y. Chen, "Stability control of autonomous ground vehicles using control-dependent barrier functions," in *IEEE Transactions on Intelligent Vehicles*, 2021, https://doi.org/10.1109/TIV.2021.3058064.

53 L. Wang, A. Ames, and M. Egerstedt, "Safety barrier certificates for collisions-free multirobot systems," *IEEE Transactions on Robotics*, vol. 33, no. 3, pp. 661–674, 2017.

54 Y. Liu, X. Wang, L. Li, S. Cheng, and Z. Chen, "A novel lane change decision-making model of autonomous vehicle based on support vector machine," *IEEE Access*, vol. 7, pp. 26543–26550, 2019.

55 A. Kurbanov, S. Grebennikov, S. Gafurov, and A. Klimchik, "Vulnerabilities in the vehicle's electronic network equipped with ADAS system," 3rd School on Dynamics of Complex Networks and their Application in Intellectual Robotics, Innopolis, Russia, 2019.

2

Vehicle, Path, and Path Tracking Models

This chapter starts with the introduction of the basic tire and longitudinal and lateral vehicle dynamics models that will be used in the rest of the book. Both independent slip and combined slip tire force models are presented. The nonlinear road load terms are linearized to obtain a simple, linearized longitudinal dynamics model. The nonlinear, linear, and linear speed varying single track vehicle models are used to characterize the lateral dynamics. The transfer functions from the steering input and yaw moment disturbance to yaw rate are derived for the linearized model with speed as a parameter. The nonlinear single-track vehicle model is also augmented with the longitudinal dynamics to formulate a more complete vehicle model with coupled longitudinal and lateral dynamics. The steady-state gain characteristic of the linearized single-track model is used to formulate the understeer gradient and study understeer and oversteer vehicles and critical and characteristic speeds. The path geometry is introduced, combining both distance to path and orientation error at a variable preview distance. The path geometry is combined with the single-track model to obtain both nonlinear and linearized versions of the path-tracking model with path curvature entering the model as a disturbance to be rejected. The computation method for path following distance and orientation errors is formulated. The simple pure pursuit and the more complicated Stanley steering control methods for path tracking are introduced and compared. A discussion of how the linearized path-tracking model developed and presented can be modified to form a path-tracking model for driving in reverse is presented as an ending note. The chapter ends with a brief summary and concluding remarks.

2.1 Tire Force Model

2.1.1 Introduction

The tire is one of the most important elements affecting the dynamics of a road vehicle. The only contact between a road vehicle and the road occurs at tire contact patches. There are four contact patches for a passenger vehicle, SUV, or pickup truck. The number of tires and hence contact patches are higher in the case of commercial vehicles like buses and larger trucks. The tire–road contact forces that are formed at these contact patches are the only way to control the dynamic motion of a road vehicle. While steering, throttle and brake actuation are used for automated driving and autonomous vehicle control, these are indirect means of actuation. The dynamics of a road vehicle for following a path or executing an evasive maneuver is actually controlled by indirectly manipulating the tire forces using the direct actuation interfaces of steering, throttle, and brake. True automated driving control of a road vehicle is not possible without a thorough knowledge

Autonomous Road Vehicle Path Planning and Tracking Control, First Edition.
Levent Güvenç, Bilin Aksun-Güvenç, Sheng Zhu, and Şükrü Yaren Gelbal.

of tire forces and tire force modeling. This is due to the fact that tire forces are nonlinear with saturation limits, are not static but have dynamic dependence including dependence on past conditions, are coupled among their different directional components and change considerably with road surface conditions due to weather effects, road material, tire material, tire structure, and tire wear.

The pneumatic tire with compressible air inside and a complicated structure consisting of rubber, belt, and reinforcement plies is hard to model and analyze accurately. Quite accurate finite element models exist, but these are too complicated and simulation-time-consuming for our purposes. Our approach will be to use simpler models that have found widespread use and are easy to implement in Matlab/Simulink. First, the wheel coordinate system will be explained along with the tire sideslip angle and the camber angle. Note that the tire side-slip angle is also called inclination angle in some references. Some important phenomena related to tires will be explained before presenting the math models. These are pneumatic trail, self-aligning torque, rolling radius, rolling resistance, and camber thrust. The magic tire formula will be introduced afterwards. This will be followed by extension to tire models that handle combined lateral and longitudinal forces that will be explained using simple and more complicated Dugoff and Nicolas–Comstock tire models.

2.1.2 Tire Forces/Moments and Slip

Tire forces act along the three orthogonal wheel frame axes as shown in Figure 2.1. The wheel coordinate frame *xyz* is centered at the center of the tire contact patch, and its axes are in alignment with the vehicle coordinate frame axes *XYZ* centered at the vehicle center of gravity under static conditions on even road when the steering angle is zero. The tire forces are the reaction forces between the tire and the road due to contact. Since the pneumatic tire is not a rigid body and since tire forces act in a contact zone, we are interested in their net sums integrated over the contact zone area. The longitudinal tire force F_x is along the tire x axis which is the traction/braking direction. Positive values of F_x are associated with traction or driving. Negative values of F_x are associated with braking. F_x cannot be controlled directly but is indirectly controlled by manipulating the wheel slip in the longitudinal direction through the application of throttle or braking. The lateral tire force F_y is parallel to the wheel y axis and is responsible for steering. Similar to longitudinal direction control, we cannot control F_y directly. It is controlled indirectly by generating tire sideslip α in response to changes in the steering angle. The tire forces F_x and F_y are nonlinear with respect to the corresponding slip values, saturate at large values of slip, and are also coupled. This nonideal nature of tire forces creates significant restrictions in the actuator force combinations that can be generated and poses significant constraints that must be accommodated when designing and implementing path-following controllers. These constraints become more restrictive under the more extreme driving typical of evasive maneuvering for collision mitigating path following control.

Tire forces F_x and F_y are frictional forces on the road plane. This is one of the rare instances where we like friction forces and actually want them to be as large as possible since we cannot

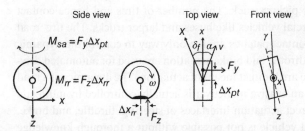

Figure 2.1 Geometry of the wheel and tire force model parameters.

control the motion of the vehicle if these friction forces go to zero in the case of driving over ice, for instance. Tire force F_z is parallel to the tire z axis and supports the part of the weight acting on that tire. The z direction corresponds to ride motion of the vehicle. Suspension systems are designed to make sure that the tire presses against the road at sufficiently high values. This is also used to characterize road holding or handling of the vehicle. The longitudinal and lateral tire forces F_x and F_y also depend on the vertical tire force F_z, with smaller F_z values resulting in smaller F_x and F_y values. In the limit as the vertical tire force F_z goes to zero, tire forces F_x and F_y also vanish. The limiting case is F_z becoming zero which means that the tire has lost all contact with the road, i.e. it has lifted off the ground. When this is true for all tires, there is no way of controlling the dynamics of a road vehicle as it is temporarily in flight. A similar situation arises when tire forces on all tires go to zero or are very small as in driving on very smooth ice. The moments in Figure 2.1 are due to tire forces F_y and F_z having a parallel offset from the tire y and z axes, respectively.

The longitudinal tire force F_x is present whenever there is longitudinal slip, i.e. whenever the tire is sliding and rolling instead of only pure rolling along the longitudinal direction. Pure rolling of a pneumatic tire almost never occurs in practice. If you take a purely rolling wheel and tire in free space, i.e. no contact with the road, and slowly start to press it against the road, the pure rolling motion will start to be a combination of sliding and rolling. The amount of sliding will increase as the tire is pressed harder against the road. The lateral tire force F_y is present whenever the wheel velocity v at the center of the wheel coordinate frame is in a different direction than the wheel longitudinal x axis. The sideways motion that results is called sideslip and the angle between the tire center velocity vector v and the x axis is called the sideslip angle which is denoted by α here. A nonzero sideslip angle α occurs during steering and also during sideways skidding of a vehicle as in drifting. The normal force F_z is always present as long as there is contact between the tire and the road. An easy approximation for it will be $Mg/4$ on level road, where M is the vehicle mass and g is the acceleration of gravity. F_z will vary between the front and rear tires due to the unsymmetric weight distribution. Weight transfer due to nonlevel road and dynamic weight transfer during throttle- or braking-based vehicle pitch motion and during cornering-based vehicle roll motion will also change F_z.

In a stationary tire, the pressure distribution in the tire contact patch in the vertical direction is symmetrical. In contrast, the pressure distribution is not symmetrical with respect to the vertical z axis for a rolling tire. Pressures are higher on the part of the tire that engages the road first as illustrated in Figure 2.2. This results in the effective vertical or normal tire force F_z not coinciding with the z axis as shown in Figure 2.2. This gives rise to the rolling resistance moment $M_{rr} = F_z \Delta x_{rr}$ about the y axis that opposes the tractive moment. ω is the tire rotational speed and v_x is the tire/wheel center speed in the x direction. $R = v_x/\omega$ is the rolling radius of the tire, i.e. the radius that the tire would have if it were in pure rolling motion without any slip in its longitudinal direction.

Just like tire force F_z is displaced by distance Δx_{rr} to the front of the tire z axis, the lateral tire force F_y is displaced by the distance Δx_{pt} to the front of the y axis as shown in Figure 2.3. The

Figure 2.2 Unsymmetrical pressure distribution in a rolling tire and rolling resistance.

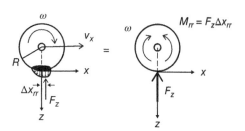

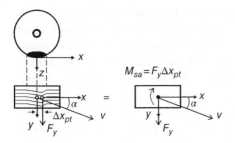

Figure 2.3 Deformation in contact zone due to lateral force and self-aligning moment.

displacement Δx_{pt} of F_y from the tire y axis on the road projection is called the pneumatic trail. The presence of the pneumatic trail gives rise to a moment about the tire z axis equal to $F_y \Delta x_{pt}$. This moment denoted by M_{sa} is called the self-aligning moment as it tries to align the tire (x direction) with its heading (direction of tire speed v), in effect trying to reduce the tire sideslip angle α. The tire contact zone deformation due to sideslip angle α is shown in more detail in Figure 2.3. The lateral deformation in the tire contact zone is due to sideslip. Drivers experience the resistance due to the self-aligning moment while they are steering as a restoring moment which tries to return the steering wheel into its neutral position. The self-aligning moment always tries to maintain zero sideslip angle. It tries to align the tire heading direction with the tire direction of travel.

Tire force characteristics are determined experimentally by tire manufacturers, using tire-testing apparatus. The test apparatus consists of a large drum which is pressed against the wheel at a chosen normal force. Necessary angles and moments can be measured and recorded during the testing. The rolling drum acts as the road for the tire. Measurements on tire-testing apparatus show that the cornering or lateral force F_y is essentially linear with side-slip angle α for small values up to about 8°. The self-aligning moment M_{sa} is less linear with α but its contribution to yaw moment is quite small. Thus, it is usually neglected in vehicle dynamics models used for controls purposes. M_{sa} is important for steering mechanism calculations and power steering control. The lateral tire force F_y is sensitive to normal load F_z, but this effect is small for the load levels of $\sim Mg/4$ that individual tires support.

Longitudinal slip or just slip in short is a measure of the amount of sliding at the tire road interface and is denoted by s. Absolute longitudinal slip is defines as

$$s_{absolute} = \omega - \omega_0 \tag{2.1}$$

and is the difference between the wheel angular speed ω and the angular speed ω_0 that would exist in the case of pure rolling. Noting that $\omega_0 = v_x/R$ in pure rolling, Eq. (2.1) becomes

$$s_{absolute} = \frac{\omega R - v_x}{R}. \tag{2.2}$$

Since absolute slip can have very large values, slip is always normalized by dividing by ω_0 to obtain the relative longitudinal slip

$$s = \frac{\omega - \omega_0}{\omega_0} = \frac{\omega - v_x/R}{v_x/R} = \frac{\omega R - v_x}{v_x}. \tag{2.3}$$

With this definition, $s \to -1$ during the limit of braking for a nonrotating wheel, i.e. in the case of a wheel lock-up ($\omega = 0$, $v_x \neq 0$). Similarly, this definition leads to $s \to \infty$ during loss of traction when the wheel rotates without any longitudinal translation of its center (ω large, $v_x = 0$) which is a spinning wheel meaning loss of traction. The range of relative longitudinal slip $-1 < s < \infty$ is not symmetrical with respect to the zero slip point. This is not desirable for control purposes, and we use the slightly modified formula given by

$$s = \frac{\omega R - v_x}{\max(v_x, \omega R, \varepsilon)} \qquad (2.4)$$

for relative longitudinal slip with the range $-1 < s < 1$. The new variable ε in the denominator is a very small number that is used to make sure that the quotient does not blow up when both v_x and ω are zero, which happens, for instance, when a vehicle starts moving after being stationary or when the vehicle comes to a full stop. This avoids a division by zero error during simulation. The negative range $-1 < s < 0$ corresponds to braking and the positive range $0 < s < 1$ corresponds to traction. Longitudinal slip is also expressed using percent slip ratio which is given as percentage of slip with 100% meaning $s = 1$ and -100% meaning $s = -1$. Percent slip ratio is given by

$$\%slip\ ratio = 100\frac{\omega R - v_x}{\max(v_x, \omega R, \varepsilon)}. \qquad (2.5)$$

2.1.3 Longitudinal Tire Force Modeling

Longitudinal tire force F_x is mainly a function of longitudinal relative slip s. Note that it also depends on sideslip angle α, road surface adhesion factor μ, and vertical tire force component F_z. Hence, it is more thorough to express the longitudinal tire force as $F_x = F_x(s, \alpha, \mu, F_z)$. A typical longitudinal tire force versus slip ratio plot for $F_x = F_x(s)$ is displayed in Figure 2.4 where the horizontal axis has slip ratio. In contrast to the lateral force and self-aligning moment plots to be presented later, there is a pronounced peak in the longitudinal force plots in Figure 2.4. For small amounts of slip, there is a linear region in the plot, where the longitudinal tire force F_x depends linearly on slip. The slope of the plot in the linear region is constant and is not affected by the road adhesion coefficient. However, the linear region becomes smaller as the road adhesion coefficient

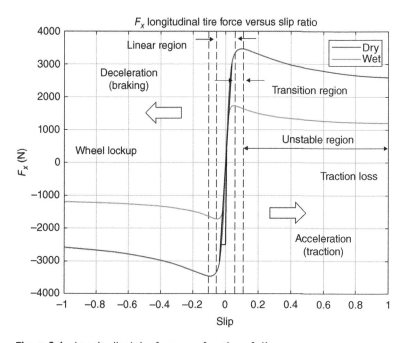

Figure 2.4 Longitudinal tire force as a function of slip.

decreases. There is a small transition region between the edge of the linear region and the peak of F_x. As slip increases beyond the transition region, the longitudinal tire force enters the unstable region, where increasing the throttle input will result in increased slip which, contrary to what is expected, rapidly decreases longitudinal force F_x until traction loss (spinning wheel) at $s = 1$ or a slip ratio of 100%. A similar phenomenon occurs in the braking direction for negative slip after the negative peak in braking force. As more braking is applied at that point increasing slip, braking force F_x decreases rapidly until wheel lock-up at $s = -1$ or a slip ratio of -100%.

Within the linear region in Figure 2.4, the linearized longitudinal tire force model is

$$F_x(s) = C_x s, \tag{2.6}$$

where the longitudinal tire stiffness coefficient C_x is the slope of the longitudinal force plot in the linear region in Figure 2.4. The empirical magic tire formula has found widespread use in the literature on tire modeling as it can easily model the shape of the longitudinal and lateral tire forces and the self-aligning moment [1]. The magic tire formula

$$F_x(s) = D \sin \left(C \tan^{-1} \phi \right) + S_v$$
$$\phi \equiv B(1 - E) \left(100s + S_h \right) + E \tan^{-1} \left(B \left(100s + S_h \right) \right) \tag{2.7}$$

is used to model the relation between longitudinal tire force and slip ratio. B, C, D, and E are empirically determined coefficients and have no physical meaning. S_h and S_v are offsets used to shift the origin of the longitudinal force plot and are both taken as zero here. In Eq. (2.7), coefficient D is the maximum value of F_x apart from the effect of S_v. B is the stiffness factor, C is the shape factor, and E is the curvature factor. The product $100BCD$ is the slope for $s + S_h = 0$ or the longitudinal tire stiffness coefficient C_x (N per unit slip) in the linear model (2.6). The independent variable s in Eq. (2.6) is slip ratio and ranges from -1 to 1. The model in Eq. (2.7) and the corresponding equation that will be presented in Sections 2.1.4 and 2.1.5 for lateral force is called the BNP (Bakker–Nyborg–Pacejka) magic formula (MF) tire force model in the literature [2].

The values of the six coefficients B, C, D, E, S_h, and S_v in Eq. (2.7) are expressed as functions of 11 new coefficients b_i, which are characteristic of a specific tire but also depend on road conditions and speed and are given by

$$C = b_0, \quad D = \mu_p F_z$$
$$\mu_p = b_1 F_z + b_2$$
$$BCD = \left(b_3 F_z^2 + b_4 F_z \right) e^{-b_5 F_z}$$
$$E = b_6 F_z^2 + b_7 F_z + b_8$$
$$S_h = b_9 F_z + b_{10}, \quad S_v = 0, \tag{2.8}$$

where $b_9 = b_{10} = 0$ is used for setting $S_h = 0$. The vertical force F_z in Eq. (2.8) is in kN so if F_z is taken as Mg or $Mg/4$ in N, it should be divided by 1000. μ_p is used as an indication of road adhesion coefficient but unlike μ which is usually between 0 and 1, it may be a very large value on the order of 1000μ. The Matlab® m-file used to generate the magic tire longitudinal forces in Figure 2.4 is given in Figure 2.5. The magic tire formula parameters used are also listed in this m-file.

2.1.4 Lateral Tire Force Modeling

Lateral tire force F_y is mainly a function of tire side slip α. It also depends on longitudinal slip s, camber angle γ, road surface adhesion factor μ, and vertical tire force component F_z. So the lateral tire force is expressed as $F_y = F_y(\alpha, s, \gamma, \mu, F_z)$ in general. A typical lateral tire force versus sideslip

```
%%%%%%%%%%%%%%%%%%%%%%%%%%%%%%%%%%%%%%%%
%
%   BNP MF Longitudinal Tire Force Plot
%
%   Autonomous Road Vehicle Path Planning and Tracking Control
%
%%%%%%%%%%%%%%%%%%%%%%%%%%%%%%%%%%%%%%%%
clear,clc
% Define percent slip range as inputs
s=[-1:0.01:1];
% Define BNP MF Fx formula parameters
b0=1.57;b1=-48;b2=1338;b3=6.8;b4=444;b5=0;
b6=0.0034;b7=-0.008; b8=0.66;b9=0;b10=0;
% Define vehicle parameters
m=1175;g=9.81;Fz=m*g/4/1000;
% Define other MF parameters
mup=b1*Fz+b2;C=b0;D=mup*Fz;BCD=(b3*Fz^2+b4*Fz)*exp(-b5*Fz);
E=b6*Fz^2+b7*Fz+b8;Sh=b9*Fz+b10;Sv=0;B=BCD/D/C;
% Calculate Fx longitudinal force
Fx_magic=D*sin(C*atan(B*(100*s+Sh)-E*(B*(100*s+Sh)-atan(B*(100*s+Sh)))))+Sv;
% Plot results
figure(1)
plot(s,Fx_magic,'b','LineWidth',2)
xlabel('slip'),ylabel('F_x (N)')
title('Fx Longitudinal Tire Force vs Slip Ratio')
grid on,hold on
% Plot results for smaller mu
mup=mup/2;
D=mup*Fz;B=BCD/D/C;
Fx_magic=D*sin(C*atan(B*(100*s+Sh)-E*(B*(100*s+Sh)-atan(B*(100*s+Sh)))))+Sv;
plot(s,Fx_magic,'r','LineWidth',2)
hold off
legend('dry','wet')
```

Figure 2.5 m-File for generating longitudinal magic tire force.

angle plot for $F_y = F_y(\alpha)$ is shown in Figure 2.6. There is a linear region in the F_y versus α plot. After a small transition region, lateral force saturates such that changes in sideslip α obtained by changing the steering angle will not significantly change lateral tire force. The lateral tire force may even decrease slightly as sideslip increases in this saturation region. The lateral tire force is the main direct actuation means for changing the orientation of the vehicle necessary for following a path. Hence, this saturating type nonlinearity is a major limitation and path-tracking controllers should be designed to operate within the linear region in Figure 2.6. Extreme driving or driving close to the envelope of safe driving will require longitudinal and lateral forces to be at the edge of the linear region and sometimes in the transition region in Figures 2.4 and 2.6, where the control designer has to be more careful as the tire is very close to the unstable region in the longitudinal direction and the saturation direction in the lateral direction. Saturating a tire force will cause integrator wind-up problems when integral control action is used in trajectory control of an autonomous road vehicle.

Within the linear region in Figure 2.6, the linearized lateral tire force model is

$$F_y(\alpha) = C_y\alpha, \tag{2.9}$$

where the cornering stiffness coefficient C_y is the slope of the lateral force plot in the linear region in Figure 2.6. An extra component of the lateral force modeled linearly is added to Eq. (2.9), resulting in

$$F_y(\alpha) = C_y\alpha - K_{ct}\gamma, \tag{2.10}$$

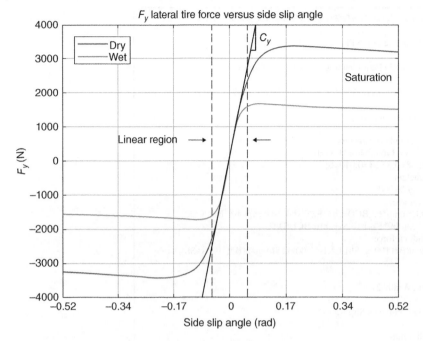

Figure 2.6 Lateral tire force as a function of sideslip angle.

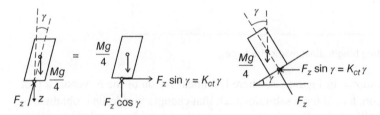

Figure 2.7 Camber thrust.

where K_{ct} is the camber thrust coefficient. The camber thrust lateral tire force $K_{ct}\gamma$ occurs due to tilting of the wheel in the presence of camber angle or tilting of the normal direction at the tire contact zone due to laterally uneven road. Both of these effects are illustrated in Figure 2.7.

The magic tire formula given by

$$F_y(\alpha) = D \sin(C \tan^{-1} \phi_\alpha) + S_v$$
$$\phi_\alpha \equiv B(1 - E)\left(\frac{180}{\pi}\alpha + S_h\right) + E \tan^{-1}\left(B\left(\frac{180}{\pi}\alpha + S_h\right)\right) \quad (2.11)$$

is used to model the relation between lateral tire force and sideslip angle α in radians. B, C, D, and E are empirically determined coefficients and have no physical meaning. S_h and S_v are offsets used to shift the origin of the lateral force plot and are both taken as zero here. In Eq. (2.11), coefficient D is the maximum value of F_y apart from the effect of S_v. B is the stiffness factor, C is the shape factor, and E is the curvature factor. The product $180BCD/\pi$ is the slope for $s + S_h = 0$ or the cornering stiffness coefficient C_y (N/rad) in the linear model (2.9). Note that while we used the same variable names

B, C, D, E, S_h, and S_v in the magic tire formulas for both the longitudinal and lateral forces, their values and formulas are different. For the lateral force F_y, the values of the six coefficients B, C, D, E, S_h, and S_v in Eq. (2.11) are expressed as functions of 14 new coefficients a_i which are not only characteristic of a specific tire but also depend on road conditions and speed and are given by

$$C = a_0, \quad D = \mu_{y_p} F_z, \quad \mu_{y_p} = a_1 F_z + a_2$$

$$BCD = a_3 \sin\left[2 \arctan\left(\frac{F_z}{a_4}\right)\right](1 - a_5|\gamma|)$$

$$E = a_6 F_z + a_7$$

$$S_h = a_8 \gamma + a_9 F_z + a_{10}$$

$$S_v = a_{11}\gamma F_z + a_{12}F_z + a_{13}$$

$$a_{11} = a_{111}F_z + a_{112}, \tag{2.12}$$

where the presence of camber angle γ can give rise to a lateral force even when sideslip α is zero. As before, the vertical force F_z in Eq. (2.12) is in kN so if F_z is taken as Mg or $Mg/4$ in N, it should be divided by 1000, and μ_p a very large value on the order of 1000μ. The Matlab m-file used to generate the magic tire lateral forces in Figure 2.6 is given in Figure 2.8. The magic tire formula parameters used are also listed in this m-file.

```
%%%%%%%%%%%%%%%%%%%%%%%%%%%%%%%%%%%%%%
%
%   BNP MF Lateral Tire Force Plot
%
%   Autonomous Road Vehicle Path Planning and Tracking Control
%
%%%%%%%%%%%%%%%%%%%%%%%%%%%%%%%%%%%%%%
clear,clc
% Define side slip angle range in degrees
alpha=[-30:0.5:30]*pi/180;
% Define vehicle parameters
m=1175;g=9.81;Fz=m*g/4/1000;
% Define MF parameters
a0=1.35;a1=-49;a2=1316;a3=1732;a4=11;a5=0.006;a6=-0.04;a7=-0.4;a8=0.003;
a9=-0.002;a10=0.16;a111=-11;a112=0.045;a12=0.17;a13=-23.5;a11=a111*Fz+a112;
% Define other MF parameters
muyp=a1*Fz+a2;C=a0;D=muyp*Fz;E=a6*Fz+a7;gama=0.013;
BCD=a3*sin(2*atan(Fz/a4))*(1-a5*abs(gama));
Sh=a8*gama+a9*Fz+a10;Sv=a11*gama*Fz+a12*Fz+a13;B=BCD/C/D;
% Calculate Fy lateral force
Fy_magic=D*sin(C*atan(B*(180*alpha/pi+Sh)-E*(B*atan(B*(180*alpha/pi+Sh)))))+Sv;
% Plot results
figure(1),plot(alpha,Fy_magic,'b','LineWidth',2)
xlabel('Side slip angle (radians)'),ylabel('F_y (N)')
title('Fy Lateral Tire Force vs Side Slip Angle')
grid on,hold on
% Plot results for smaller mu
muyp=muyp/2;D=muyp*Fz;B=BCD/D/C;
Fy_magic=D*sin(C*atan(B*(180*alpha/pi+Sh)-E*(B*atan(B*(180*alpha/pi+Sh)))))+Sv;
plot(alpha,Fy_magic,'r','LineWidth',2)
hold off,legend('dry','wet')
```

Figure 2.8 m-File for generating lateral magic tire force.

2.1.5 Self-aligning Moment Model

The self-aligning moment is modeled using the magic tire formula

$$M_{sa}(\alpha) = D \sin(C \tan^{-1} \phi_{sa}) + S_v$$
$$\phi_{sa} \equiv B(1 - E)\left(\frac{180}{\pi}\alpha + S_h\right) + E \tan^{-1}\left(B\left(\frac{180}{\pi}\alpha + S_h\right)\right), \tag{2.13}$$

where the independent variable is side slip angle α in radians. The empirically determined coefficients B, C, D, and E and the offsets S_h and S_v are expressed as functions of 18 new coefficients c_i, as

$$C = c_0, \quad D = c_1 F_z^2 + c_2 F_z,$$
$$E = \left(c_7 F_z^2 + c_8 F_z + c_9\right)(1 - c_{10}|\gamma|),$$
$$BCD = \left(c_3 F_z^2 + c_4 F_z\right)(1 - c_6|\gamma|)e^{-c_5 F_z},$$
$$S_h = c_{11}\gamma + c_{12}F_z + c_{13},$$
$$S_v = \left(c_{14}F_z^2 + c_{15}F_z\right)\gamma + c_{16}F_z + c_{17}, \tag{2.14}$$

and a typical plot is shown in Figure 2.9. In plotting Figure 2.9, the coefficients used were $c_0 = 2.46$, $c_1 = -2.77$, $c_2 = -2.9$, $c_3 = -0$, $c_4 = -3.6$, $c_5 = -0.1$, $c_6 = 0.0004$, $c_7 = 0.22$, $c_8 = -2.31$, $c_9 = 3.87$, $c_{10} = 0.0007$, $c_{11} = -0.05$, $c_{12} = -0.006$, $c_{13} = 0.33$, $c_{14} = -0.04$, $c_{15} = -0.4$, $c_{16} = 0.092$, and $c_{17} = 0.0114$ and vehicle parameters used were $M = 1175$ kg and $\gamma = 0°$. Within the linear region in Figure 2.9, self-aligning moment is well approximated by

$$M_{sa}(\alpha) = -C_{sa}\alpha, \tag{2.15}$$

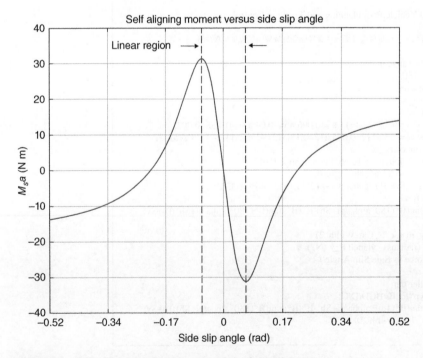

Figure 2.9 Self-aligning moment as a function of sideslip angle.

where C_{sa} is the constant of proportionality. The negative slope in the linear region in Figure 2.9 is due to the fact that the self-aligning moment M_{sa} opposes side slip α in direction within that region. Within the linear region, the pneumatic trail Δx_{pt} in Figure 2.3 increases as sideslip angle increases, thus, increasing the self-aligning moment. Outside the linear region, though, the pneumatic trail starts decreasing which also decreases the self-aligning moment.

2.1.6 Coupling of Tire Forces

In addition to their nonlinear behavior, the longitudinal and lateral tire forces are also coupled, meaning that they cannot both achieve their maximum value at the same time. The lateral and longitudinal tire forces are planar friction forces. Their resultant friction force in the tire-road interaction plane is

$$F = \sqrt{F_x^2 + F_y^2}. \tag{2.16}$$

If μ_{max} is the maximum value of the road adhesion coefficient μ, then

$$F = \sqrt{F_x^2 + F_y^2} \le \mu_{max} F_z \tag{2.17}$$

which can be rewritten to obtain the grip margin

$$grip\ margin = \mu_{max} F_z - \sqrt{F_x^2 + F_y^2}, \tag{2.18}$$

which is the distance between the current tire force resultant on the road plane and the available maximum force as illustrated in Figure 2.10. The horizontal axis shows the lateral force and the vertical axis shows the longitudinal force. The resultant force F is limited by the maximum possible planar force $\mu_{max} F_z$ and the resulting circle with that radius shown in Figure 2.10 is called Kamm's circle or the friction circle. The circle in Figure 2.10 is actually a tire force circle. The name friction circle is used as the circular envelope with radius $\mu_{max} F_z$ corresponds to the theoretical maximum grip value as shown in Figure 2.10 when the friction coefficients in the x and y directions obey

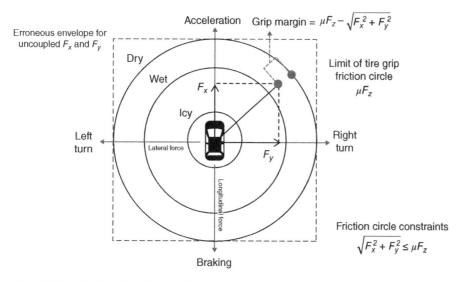

Figure 2.10 Friction circle (Kamm Circle).

$\mu_x = \mu_y = \mu$, where μ_{max} is the maximum value of the combined friction coefficient μ. The individual friction coefficients are determined by

$$F_x = \mu_x F_z \Rightarrow \mu_x \equiv \frac{F_x}{F_z}, \quad F_y = \mu_y F_z \Rightarrow \mu_y \equiv \frac{F_y}{F_z}$$

$$\mu = \frac{F}{F_z} = \frac{\sqrt{F_x^2 + F_y^2}}{F_z} = \sqrt{\left(\frac{F_x}{F_z}\right)^2 + \left(\frac{F_y}{F_z}\right)^2}$$

$$\mu = \sqrt{\mu_x^2 + \mu_y^2}. \tag{2.19}$$

For $\mu_x \neq \mu_y$, the friction circle or tire force circle becomes the friction ellipse of tire force ellipse.

For road conditions with smaller friction coefficients like wet or icy roads, the radius of the Kamm's circle and hence the available envelope of safe driving shrinks. Neglecting this coupling between the longitudinal and lateral tire force leads to the square envelope in Figure 2.10. Linearized models for tire force neglect this coupling along with the other nonlinearities of tire forces and trajectory-tracking controllers based on them may erroneously demand tire forces that are outside the circular envelope in Figure 2.10. One possible remedy will be to use limits on F_x, F_y, forcing them to lie in their linear ranges and to impose $F = \sqrt{F_x^2 + F_y^2} \leq \mu_{max} F_z$ on their resultant during the controller design. Another possible remedy which is also used very commonly is to simulate the designed control system using a higher fidelity model with accurate, nonlinear tire dynamics, and to go back and redesign if the tire saturation limits are violated.

The second approach of simulation with a realistic tire model requires the use of a tire force model that also incorporates the coupling of the longitudinal and lateral components. Extensions of the magic tire formulas presented earlier and various other models that incorporate this coupled behavior exist and are also built into many of the available vehicle dynamics simulation programs. In this section, we will introduce the simplest of these which are the two Dugoff tire force models [3]. In the simple Dugoff model, longitudinal and lateral forces are modeled similar to their linearized versions as

$$F_x = fC_x s, \tag{2.20}$$

$$F_y = fC_y \alpha, \tag{2.21}$$

with the only difference being the coupling coefficient f which is defined by

$$f = \begin{cases} 1, & F_R \leq \dfrac{\mu F_z}{2} \\ \left(2 - \dfrac{\mu F_z}{2F_R}\right) \dfrac{\mu F_z}{2F_R}, & F_R > \dfrac{\mu F_z}{2} \end{cases}, \tag{2.22}$$

where the resultant force F_R is defined as

$$F_R = \sqrt{(C_x s)^2 + (C_y \alpha)^2}. \tag{2.23}$$

The simple Dugoff model is very easy to apply as there are no parameters to adjust other than the slopes C_x and C_y of the linear tire force models and the road adhesion coefficient μ. This is also a disadvantage of the method since the shapes of the tire force plots cannot be adjusted to fit the shape of experimental data. Longitudinal and lateral tire forces obtained using the simple Dugoff tire model are shown in Figure 2.11 and exhibit the inherent coupling between these two tire force components.

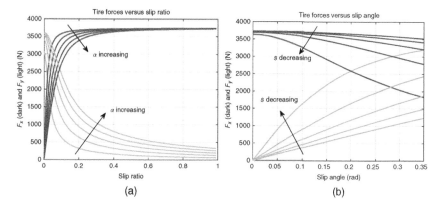

Figure 2.11 Simple Dugoff tire model forces (a) as a function of slip ratio (b) as a function of side slip angle.

There is also the second Dugoff tire model which is slightly more complicated than the first and is called complicated Dugoff model here and is governed by

$$F_x = \frac{C_x s}{1 - s} f(S),$$
(2.24)

$$F_y = \frac{C_y \tan \alpha}{1 - s} f(S),$$
(2.25)

for the computation of the longitudinal and lateral force components where the combined slip S is defined as

$$S = \frac{\mu F_z [1 - \varepsilon_r U \sqrt{s^2 + \tan^2 \alpha}](1 - s)}{2\sqrt{C_x^2 s^2 + C_y^2 \tan^2 \alpha}},$$
(2.26)

and the function $f(S)$ is calculated as

$$f(S) = \begin{cases} S(2 - S), & \text{for } S < 1 \\ 1, & \text{for } S > 1 \end{cases}.$$
(2.27)

In the complicated Dugoff model, U is speed in the longitudinal direction and ε_r is usually taken as zero.

The Dugoff models are easy to implement in simulation. The Nicolas–Comstock (NC) model is also a relatively easy to implement model that modifies the BNP MF tire model forces introduced earlier [4]. The forces in the longitudinal and lateral direction in the Nicolas–Comstock model are calculated using

$$F_x(s, \alpha) = \frac{F_x(s) F_y(\alpha) s}{\sqrt{s^2 F_y^2(\alpha) + F_x^2(s) \tan^2 \alpha}},$$
(2.28)

$$F_y(\alpha, s) = \frac{F_x(s) F_y(\alpha) \tan \alpha}{\sqrt{s^2 F_y^2(\alpha) + F_x^2(s) \tan^2 \alpha}},$$
(2.29)

where $F_x(s)$ and $F_y(\alpha)$ are the BNP MF tire forces given in Eqs. (2.7) and (2.11). The model in Eqs. (2.28) and (2.29) was later modified to obtain the Modified–Nicolas–Comstock (MNC) model given by

$$F_x(s, \alpha) = \frac{F_x(s)F_y(\alpha)s}{\sqrt{s^2 F_y^2(\alpha) + F_x^2(s)\tan^2 \alpha}} \frac{\sqrt{s^2 C_y^2 + (1-|s|)^2 F_x^2(s)\cos^2 \alpha}}{sC_y}, \tag{2.30}$$

$$F_y(\alpha, s) = \frac{F_x(s)F_y(\alpha)\tan \alpha}{\sqrt{s^2 F_y^2(\alpha) + F_x^2(s)\tan^2 \alpha}} \frac{\sqrt{(1-|s|)^2\cos^2 \alpha F_y^2(\alpha) + C_x^2 \sin^2 \alpha}}{C_x \sin \alpha}, \tag{2.31}$$

where the first part in both equations is the Nicolas–Comstock model multiplied by a second quotient which is the modification term for the MNC model and changes the shape of the friction ellipse. The MNC tire forces corresponding to the BNP model presented earlier for $F_x(s)$ and $F_y(\alpha)$ are displayed in Figure 2.12. The m-file for the MNC tire force calculation function is given in Figure 2.13. C_x and C_y are the slopes of the BNP MF $F_x(s)$ versus s and $F_y(\alpha)$ versus α plots, respectively.

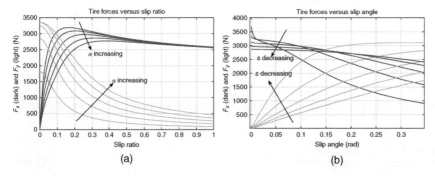

Figure 2.12 MNC tire forces (a) as a function of slip ratio (b) as a function of side slip angle.

```
%%%%%%%%%%%%%%%%%%%%%%%%%%%%%%%%%%%%%%
%
%   MNC Tire Forces Computation Function
%
%   Autonomous Road Vehicle Path Planning and Tracking Control
%
%%%%%%%%%%%%%%%%%%%%%%%%%%%%%%%%%%%%%%

% Calculate BNP model forces Fx, Fy and then run function below

function [Fx_MNC, Fy_MNC] = MNC_Model(Fx,Fy,s,alpha,Cx,Cy)
% Nicolas Comstock (NC) Model
% Takes individual Lateral & Longitidunal Forces, s and alpha values as inputs,
% gives combined Lateral & Long. forces.
Fx_MNC = (Fx*Fy*s/sqrt(s*s*Fy*Fy+Fx*Fx*tan(alpha)*tan(alpha)))...
*(sqrt(s*s*Cy*Cy+(1-abs(s))^2*Fx*Fx*cos(alpha)^2)/(s*Cy));
Fy_MNC = (Fx*Fy*tan(alpha)/sqrt(s*s*Fy*Fy+Fx*Fx*tan(alpha)*tan(alpha)))...
*(sqrt((1-abs(s))^2*cos(alpha)^2*Fy*Fy+Cx^2*sin(alpha)^2)/(Cx*sin(alpha)));
End
```

Figure 2.13 m-File for the MNC tire forces function.

There are many more tire models in the literature that were developed to capture the coupling behavior between the longitudinal and lateral directions accurately. Real-life tire forces are too complex to be modeled very accurately under all possible conditions using analytical models like the ones presented in this section. The approach taken in this book is to use the very simple linear longitudinal and lateral tire force models while paying attention to coupling effects and treating differences with actual behavior as model uncertainty to be handled by the robustness of the path-tracking controllers being designed. The designed controllers are validated first in a more realistic simulation with a higher fidelity tire force model, followed in some instances by experimental validation.

2.2 Vehicle Longitudinal Dynamics Model

The forces/torques that oppose the motion of a vehicle are known as *Road Load*. In general, the road load can be broken down into several components as follows: rolling resistance from the tires, aerodynamic drag, and road grade. Road grade load is resistive (uphill) or supportive (downhill) depending on the sign of the road slope. The longitudinal tractive force provides the inertial force required to accelerate the vehicle against the road load as illustrated in Figure 2.14. In the case of braking, the road load helps the longitudinal braking force in slowing down (decelerating) the vehicle. The direction of F_x will be reversed in Figure 2.14 in the case of braking.

The longitudinal dynamics equation for the vehicle is

$$F_x - \underbrace{\frac{1}{2}\rho_a C_d A_f (v_x + v_{wind})^2}_{\text{aerodynamic resistance}} - \underbrace{C_r(v_x)Mg \cos \theta}_{\text{rolling resistance}} - \underbrace{Mg \sin \theta}_{\text{grade resistance}} = M\frac{dv_x}{dt}, \tag{2.32}$$

where F_x is the sum of the tire longitudinal forces, M is effective mass of the vehicle including a small increase due to the rotating inertia effect, and v_x is vehicle longitudinal speed. ρ_a is the density of air, C_d is the drag coefficient, A_f is frontal area of the vehicle, v_{wind} is headwind speed, C_r is rolling resistance coefficient, and θ is road grade angle. The frontal area A_f is typically about $2\,\text{m}^2$, while the aerodynamic drag coefficient C_d is about 0.3 for passenger vehicles. The rolling resistance C_r is approximated by

$$C_r(v_x) = \underbrace{C_{r0}}_{\text{speed independent}} + \underbrace{C_{r1}v_x}_{\text{speed dependent}}, \tag{2.33}$$

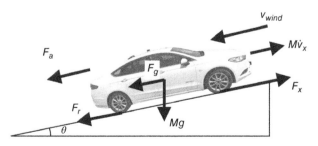

Figure 2.14 Longitudinal road load.

where a constant C_{r0} and a linearly speed-dependent part $C_{r1}v_x$ are used. The actual rolling resistance might depend nonlinearly on longitudinal speed but Eq. (2.33) is still a good approximation for a simplified analysis. The rolling resistance coefficient is quite often modeled simply as a constant term.

Note that the road load in Eq. (2.32) can be expressed as

$$F_{road\_load} \equiv \left((C_{r0} + C_{r1}v_x)Mg \cos\theta + \frac{1}{2}\rho_a C_d A_f (v_x + v_{wind})^2 \right) + Mg \sin\theta, \tag{2.34}$$

which for a flat road ($\theta = 0°$) and zero wind ($v_{wind} = 0$) becomes

$$F_{road\_load} \equiv C_{r0}Mg + (C_{r1}Mg)v_x + \left(\frac{1}{2}\rho_a C_d A_f \right) v_x^2 \equiv a + bv_x + cv_x^2, \tag{2.35}$$

$$a \equiv C_{r0}Mg, \quad b \equiv C_{r1}Mg, \quad c \equiv \frac{1}{2}\rho_a C_d A_f. \tag{2.36}$$

A coast-down test is used to determine the coefficients a, b, and c in the second order polynomial in Eq. (2.35). The transmission is switched to neutral on a flat, dry road under no wind conditions, and the speed is recorded during this test. The simplified longitudinal dynamics model of a vehicle during the coast-down test becomes

$$a + bv_x + cv_x^2 = M\frac{dv_x}{dt}. \tag{2.37}$$

Best fit values of coefficients a, b, and c are determined using regression analysis to make the collected coast-down longitudinal speed data fit Eq. (2.37) in a least squares sense. The block diagram representing the longitudinal dynamic model in Eq. (2.32) is shown in Figure 2.15.

The next part of longitudinal dynamics comes from the modeling of the tires. This is how the throttle and brake commands modify the tire longitudinal force F_x for providing traction or braking. The free body diagram of a drive wheel is shown in Figure 2.16. A drive wheel is also called an active wheel as it is connected to the prime mover (internal combustion engine or electric motor) through the transmission. The prime mover torque T_m is multiplied by drivetrain efficiency λ and transmission ratio η_t. Braking torque T_b is also shown in Figure 2.16 for the sake of completeness. Note that torques T_m and T_b are not applied simultaneously during autonomous driving of an

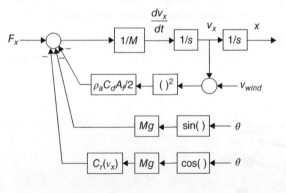

Figure 2.15 Vehicle longitudinal dynamics block diagram.

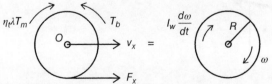

Figure 2.16 Free body diagram of drive wheel.

automated vehicle as throttle and brake actuation do not take place at the same time. However, this simultaneous application may take place during the intervention of the traction control system or the electronic stability control system. While the brake torque T_b can be present at each tire, driving or traction torque T_m does not exist for the nondrive or passive tires like the rear tires in a front wheel drive vehicle, for instance.

Taking moments about the wheel center O in Figure 2.16 results in

$$\sum M_O : \ \eta_t \lambda T_m - T_b - F_x(s)R = I_w \frac{d\omega}{dt}, \tag{2.38}$$

where I_w is the wheel and tire moment of inertia.

The block diagram corresponding to the wheel dynamic model in Eq. (2.38) is shown in Figure 2.17. The overall one-wheel block diagram in Figure 2.18 is obtained when longitudinal dynamics and cruise control are added. The block diagram in Figure 2.18 can be extended to have all four-wheel dynamic models that generate a total longitudinal tractive or brake force combined with the vehicle longitudinal dynamics. This is not shown here as the block diagram becomes too large. Two front drive wheels and two rear passive wheels would be used in that case for a front-wheel drive vehicle, with minor changes being needed for different drive configurations like rear wheel drive or all-wheel drive. The powertrain model and the brake dynamic models are not

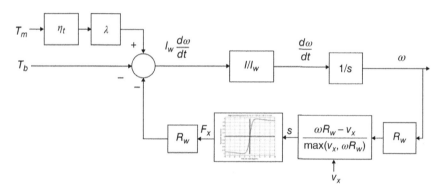

Figure 2.17 Block diagram of wheel dynamics.

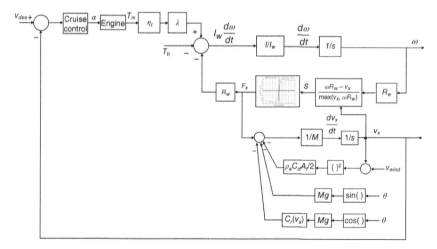

Figure 2.18 Longitudinal dynamics combined with one wheel model.

shown in Figure 2.18 and would need to be added for designing a cruise control system. Linearizing some of the nonlinearities and neglecting some of them in Figure 2.18, the simple longitudinal dynamics models from longitudinal traction and brake actuation torques to longitudinal speed are obtained as

$$v_x = \frac{K(v_x)e^{-\tau_d s}}{\tau(v_x)s + 1} T_m + \frac{K_b(v_x)e^{-\tau_{db}s}}{\tau_b(v_x)s + 1} T_b \tag{2.39}$$

or its modified version

$$v_x = \frac{K(v_x)e^{-\tau_d s}}{\tau(v_x)s + 1} \left(1 + W_{Tm}\Delta_m\right) T_m + \frac{K_b(v_x)e^{-\tau_{db}s}}{\tau_b(v_x)s + 1} \left(1 + W_{Tb}\Delta_b\right) T_b + d, \tag{2.40}$$

where $W_{Tm}\Delta_m$ ($\|\Delta_m\|_\infty \leq 1$) and $W_{Tb}\Delta_b$ ($\|\Delta_b\|_\infty \leq 1$) are unstructured uncertainty representing unmodeled dynamics and d represents both measurable disturbances like road grade and also more arbitrary disturbances like wind. The time delays in Eq. (2.39) do not come from the block diagram in Figure 2.18 but are added to model the delay in powertrain combustion and transmission delay with τ_d and the brake actuation delay with τ_{db}. τ and τ_b represent the time constants for traction and braking actuation, respectively. K and K_b are the static gains from traction and brake torque to longitudinal speed, respectively. The time constants and gains in Eq. (2.39) vary with the operational longitudinal speed. The time delays, on the other hand, vary more arbitrarily and lower and upper bounds are used to capture their uncertainty range. Adjusting these gains and time constants, Eq. (2.39) can also be expressed using throttle angle (accelerator pedal position) and brake angle (brake pedal position) as inputs and longitudinal speed as output. Note that drive and brake torques are not applied simultaneously in Eq. (2.39) in automated driving unless individual wheel braking is used for torque vectoring, traction control intervention, or electronic stability control intervention in the simultaneous presence of throttle input. The nonlinear longitudinal vehicle dynamics with the speed dependence of even its simplified linearized version make cruise control of a vehicle throughout all of its possible speed range from zero to maximum speed quite difficult. Note that current cruise control systems in series produced vehicles typically work only at highway speeds. In autonomous driving, however, a cruise controller also has to work at low speeds. This requires the use of a finely tuned, speed scheduled controller for longitudinal speed (cruise) control. Low-speed cruise control of an internal combustion engine powered vehicle is quite difficult. It is relatively much easier to control low-speed cruise for fully electric and hybrid electric vehicles since low-speed control of electric motors is relatively easier. It is customary to design the speed scheduled cruise controller using the simple linearized model of longitudinal dynamics and to later check performance using both the nonlinear model and a higher fidelity model.

2.3 Vehicle Lateral Dynamics Model

2.3.1 Geometry of Cornering

Lateral dynamics modeling is the most important part of building a path-tracking model for an autonomous vehicle. The well-known single-track model [5] will be developed and used here as it is the simplest model which accurately captures the handling behavior of a road vehicle. We will start with a discussion of the geometry of turning as illustrated in Figure 2.19 for a front wheel steering (FWS) vehicle. This is a single-track representation of the vehicle where the two front tires and the two rear tires have been lumped into one front tire and one rear tire, respectively. The

Figure 2.19 Geometry of high-speed cornering.

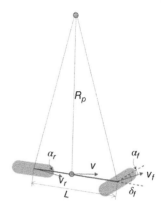

steering angle δ_f is approximated by

$$\delta_f - \alpha_f + \alpha_r = \frac{L}{R_p} \Rightarrow \delta_f = \frac{L}{R_p} + \alpha_f - \alpha_r, \tag{2.41}$$

where α_f and α_r are the front and rear tire sideslip angles, R_p is the instantaneous radius of curvature of the path being followed, and L is the wheelbase of the vehicle, i.e. distance between its front and rear axles. The yaw rate r of the vehicle is

$$r \equiv \frac{d\psi}{dt} \cong \frac{V}{R_p}, \tag{2.42}$$

where ψ is the yaw angle and V is the vehicle velocity at its center of mass.

High-speed cornering as shown in Figure 2.19 corresponds to large values of the tire side slips α_f and α_r. Recall from the discussion of tire forces that large tire sideslip values mean that longitudinal tire forces will be limited in value, limiting traction or braking capability during cornering. In the case of low-speed cornering which corresponds to nonextreme driving, the geometry of turning is simplified as shown in Figure 2.20 with small tire side slips, where $\alpha_f = \alpha_r = 0°$ can be used as an approximation. In this case, Eq. (2.41) becomes

$$\delta_f = \frac{L}{R_p} \tag{2.43}$$

which is called neutral steering and is characterized by the more general condition $\alpha_f = \alpha_r$. This means that there is no relative sideslip between the front and rear tires. This also corresponds to a

Figure 2.20 Geometry of low-speed cornering.

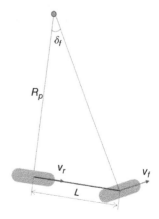

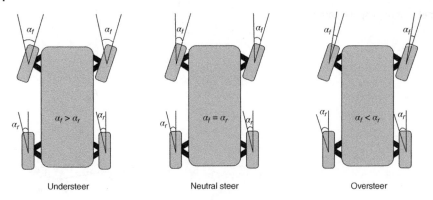

Understeer   Neutral steer   Oversteer

Figure 2.21 Illustration of understeer and oversteer.

vehicle that can perfectly track a circular path at the steering angle given by Eq. (2.43). There are two other possibilities which are discussed next.

When $\alpha_f > \alpha_r$, the front tires are slipping more in the sideways direction as compared to the rear tires and this condition is called understeer that is illustrated in Figure 2.21. An understeer vehicle will keep going forward during cornering, i.e. the vehicle turns less sharply than the steering wheel input. Understeer is safer as compared to oversteer since the speed of the vehicle can easily be decreased to regain control. A vehicle exhibiting larger amounts of understeer is sometimes said to *plow* through the turn. An understeering vehicle requires the autonomous driving system to increase steering wheel input or reduce speed or do both during cornering. This condition is easy to control for a driver, so all passenger cars and light trucks are designed with a small amount of understeer.

On the other hand, when $\alpha_f < \alpha_r$ as shown in Figure 2.21, the rear tires are slipping more in the sideways direction as compared to the front tires, resulting in oversteering. In an oversteer vehicle, the rear tires lose grip (larger sideslip) and the back of the vehicle slides outwards of the curve during cornering. Since this is difficult to handle by a driver, oversteer is not a desirable characteristic. It should be noted that the automated path-tracking controller of an autonomous vehicle should be able to handle both understeer and oversteer characteristics as a properly designed robust feedback control system will take care of the deviations from neutral steering. Both understeer and oversteer introduce uneven steering actuation as compared to neutral steering and will have an effect on path-tracking controller design performance. This will be more important when a simple geometry/kinematics-based path-tracking controller is used. While these types of simple steering controllers work well for mobile robots operating at low speeds, they should not be employed for autonomous road vehicles, especially during higher-speed driving. For best performance, steering angle compensation and speed manipulation should be used simultaneously for understeer or oversteer vehicles. Note that understeer and oversteer characteristics depend on the longitudinal weight distribution of the vehicle, i.e. whether the front or the rear axle support more of the weight of the vehicle. Understeer/oversteer also depend on the cornering stiffness and road adhesion coefficient differences between the front and rear tires.

2.3.2 Single-Track Lateral Vehicle Model

The single-track vehicle model is illustrated in Figure 2.22. OXY is the inertial reference frame that is fixed on the road such that the center of mass *CG* of the vehicle coincides with its center *O* in the

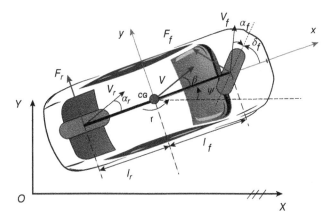

Figure 2.22 Geometry of single-track vehicle model.

beginning of motion at time $t = 0$ seconds. The coordinate frame $CGxy$ is fixed on the vehicle with x and y being the longitudinal and lateral directions, respectively. The vehicle center of mass velocity is V at vehicle (chassis) sideslip angle of β with respect to the positive x axis, measured counterclockwise positive. The tire/wheel center points have velocities of V_f and V_r with tire sideslip angles of α_f and α_r for the front and rear tires, respectively. The vehicle in Figure 2.22 is FWS with δ_f being the steering angle. It is very easy to add rear wheel steering (RWS) for a four-wheel-steering system by adding steering angle δ_r. l_f and l_r are the distances from the center of mass CG to the front and rear axles, constrained by $l_f + l_r = L$, where L is the wheelbase. Only the lateral tire forces $F_f(\alpha_f)$ and $F_r(\alpha_r)$ are considered in the conventional single-track model. The other assumptions are that vehicle motion takes place in the OXY plane and pitch and roll rotational motions are neglected. So the vehicle can only translate along the x and y directions and rotate about its vertical z axis with yaw rate r. Hence, this is a three geometric degrees-of-freedom (two translations and one rotation) model.

The acceleration $\mathbf{a}$ of the center of mass of the vehicle is obtained by differentiating its velocity vector $\mathbf{V}$ as

$$\mathbf{a} = \left(\frac{d\mathbf{V}}{dt}\right)_{OXY} = \left(\frac{d\mathbf{V}}{dt}\right)_{CGxy} + r\mathbf{k}x\mathbf{V} = (a_x\mathbf{i} + a_y\mathbf{j}) + r\mathbf{k}x\mathbf{V}, \tag{2.44}$$

where the first term is the acceleration seen by an observer in the vehicle and the second term is the change in direction of the vehicle velocity vector as it rotates with yaw rate $r\mathbf{k}$ with respect to the fixed coordinate frame. Evaluating the cross product in Eq. (2.44) and re-arranging results in

$$\mathbf{a} = a_x\mathbf{i} + a_y\mathbf{j} + r\mathbf{k}x(V_x\mathbf{i} + V_y\mathbf{j}) = (a_x - rV_y)\mathbf{i} + (a_y + rV_x)\mathbf{j}. \tag{2.45}$$

Applying Newton's second law to the free body diagram with tire lateral forces F_f and F_r in Figure 2.22 results in

$$\sum F_x = M(a_x - rV_y) = -F_f(\alpha_f)\sin \delta_f - F_r(\alpha_r)\sin \delta_r,$$
$$\sum F_y = M(a_y + rV_x) = F_f(\alpha_f)\cos \delta_f + F_r(\alpha_r)\cos \delta_r, \tag{2.46}$$

where a possible RWS angle δ_r has also been included. The moment equation about CG in Figure 2.22 results in

$$\sum M_z = I_z\ddot{\psi} = I_z\dot{r} = l_f \cos \delta_f F_f(\alpha_f) - l_r \cos \delta_r F_r(\alpha_r) + M_{zd}, \tag{2.47}$$

where M_{zd} is a yaw disturbance moment, and I_z is its moment of inertia, both about the vehicle z axis at CG. The left- and right-hand sides of Eqs. (2.46) and (2.47) can be written more compactly as

$$\begin{bmatrix} \sum F_x \\ \sum F_y \\ \sum M_z \end{bmatrix} = \begin{bmatrix} -\sin \delta_f & -\sin \delta_r \\ \cos \delta_f & \cos \delta_r \\ l_f \cos \delta_f & -l_r \cos \delta_r \end{bmatrix} \begin{bmatrix} F_f(\alpha_f) \\ F_r(\alpha_r) \end{bmatrix} \equiv \mathbf{T_{sp}}(\delta_f, \delta_r) \begin{bmatrix} F_f(\alpha_f) \\ F_r(\alpha_r) \end{bmatrix}, \tag{2.48}$$

where $\mathbf{T_{sp}}$ is called the *steering projection* as it projects the tire lateral forces along the vehicle longitudinal (x) and lateral (y) axes. Equations (2.46)–(2.48) can be combined to obtain

$$\begin{bmatrix} \sum F_x \\ \sum F_y \\ \sum M_z \end{bmatrix} = \begin{bmatrix} M & 0 & 0 \\ 0 & M & 0 \\ 0 & 0 & I_z \end{bmatrix} \begin{bmatrix} a_x \\ a_y \\ \dot{r} \end{bmatrix} + \begin{bmatrix} -V_y \\ V_x \\ 0 \end{bmatrix} r = \mathbf{T_{sp}}(\delta_f, \delta_r) \begin{bmatrix} F_f(\alpha_f) \\ F_r(\alpha_r) \end{bmatrix} \tag{2.49}$$

or the main dynamics equations

$$\begin{bmatrix} a_x \\ a_y \\ \dot{r} \end{bmatrix} = \frac{d}{dt} \begin{bmatrix} V_x \\ V_y \\ r \end{bmatrix} = \begin{bmatrix} \frac{1}{M} & 0 & 0 \\ 0 & \frac{1}{M} & 0 \\ 0 & 0 & \frac{1}{I_z} \end{bmatrix} \left(\begin{bmatrix} V_y \\ -V_x \\ 0 \end{bmatrix} r + \mathbf{T_{sp}}(\delta_f, \delta_r) \begin{bmatrix} F_f(\delta_f - \beta_f) \\ F_r(\delta_r - \beta_r) \end{bmatrix} \right), \tag{2.50}$$

where $\alpha_f = \delta_f - \beta_f$ and $\alpha_r = \delta_r - \beta_r$ were used. Integration of the left-hand side of Eq. (2.50) will result in vehicle center of mass velocity components V_x and V_y and yaw rate r. These three variables can equivalently be represented by the commonly used single-track model state variables yaw rate r, vehicle center of mass speed V, and chassis sideslip angle β with the two latter ones related to V_x and V_y as

$$V = \sqrt{V_x^2 + V_y^2}, \quad \beta = \tan^{-1}\left(\frac{V_y}{V_x}\right). \tag{2.51}$$

Note that β_f and β_r in Eq. (2.50) are given by

$$\beta_f = \tan^{-1}\left(\frac{V_{fy}}{V_{fx}}\right) = \tan^{-1}\left(\frac{V \sin \beta + rl_f}{V \cos \beta}\right) = \tan^{-1}\left(\tan \beta + \frac{rl_f}{V \cos \beta}\right), \tag{2.52}$$

$$\beta_r = \tan^{-1}\left(\frac{V_{ry}}{V_{rx}}\right) = \tan^{-1}\left(\frac{V \sin \beta - rl_r}{V \cos \beta}\right) = \tan^{-1}\left(\tan \beta - \frac{rl_r}{V \cos \beta}\right) \tag{2.53}$$

using basic kinematics. Equations (2.50), (2.52), and (2.53) represent the dynamics of the nonlinear single-track model. Equations (2.52) and (2.53) represent kinematics/geometry equations and can be written compactly as

$$\begin{bmatrix} \beta_f \\ \beta_r \end{bmatrix} = \mathbf{f}_{k/g}(\beta, V, r). \tag{2.54}$$

The block diagram representation of the nonlinear single-track vehicle model is shown in Figure 2.23. δ_f for a FWS and possibly δ_r for a RWS vehicle are the inputs to this model and β, V, and r are the outputs. The kinematics/geometry relation in Eq. (2.54) result in the front and rear tire center velocity angles β_f and β_r which are subtracted from the input steering angles to obtain the tire sideslip angles. The tire sideslip angles are inputs to the front and rear tire lateral force models. The tire lateral forces enter the steering projection $\mathbf{T_{sp}}$ and are transformed into the x, y directions and a moment about the z axis. These enter the vehicle dynamics block where the dynamics equations are solved and integrated to obtain the three state variables of β, V, and r which are also inputs into the kinematics/geometry block.

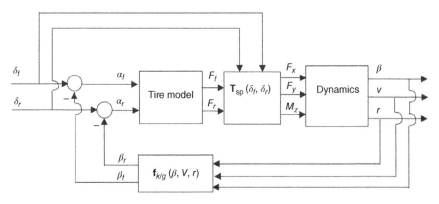

Figure 2.23 Block diagram of nonlinear single-track vehicle model.

It is also possible to express the nonlinear single-track vehicle model in the equivalent form

$$
\begin{bmatrix} Mv(\dot{\beta} + r) \\ M\dot{V} \\ I_z\dot{r} \end{bmatrix} = \begin{bmatrix} (F_f \sin \delta_f \sin \beta) + (F_f \cos \delta_f + F_r)\cos \beta \\ -F_f \sin \delta_f \cos \beta + (F_f \cos \delta_f + F_r)\sin \beta \\ l_f F_f \cos \delta_f - l_r F_r \end{bmatrix} + \begin{bmatrix} 0 \\ 0 \\ 1 \end{bmatrix} M_{zd} \tag{2.55}
$$

or

$$
\begin{bmatrix} \dot{\beta} \\ \dot{V} \\ \dot{r} \end{bmatrix} = \begin{bmatrix} -1 \\ 0 \\ 0 \end{bmatrix} r + \begin{bmatrix} ((F_f \sin \delta_f \sin \beta) + (F_f \cos \delta_f + F_r)\cos \beta)/MV \\ (-F_f \sin \delta_f \cos \beta + (F_f \cos \delta_f + F_r)\sin \beta)/M \\ (l_f F_f \cos \delta_f - l_r F_r)/I_z \end{bmatrix} + \begin{bmatrix} 0 \\ 0 \\ 1/I_z \end{bmatrix} M_{zd} \tag{2.56}
$$

$$
\begin{bmatrix} \dot{\beta} \\ \dot{V} \\ \dot{r} \end{bmatrix} = \mathbf{f}_{\mathrm{NLst}}(\beta, V, r) + \begin{bmatrix} 0 \\ 0 \\ 1/I_z \end{bmatrix} M_{zd}, \tag{2.57}
$$

where

$$
F_f = F_f(\alpha_f) = F_f\left(\delta_f - \tan^{-1}\left(\tan \beta + \frac{l_f r}{V} \right) \right), \tag{2.58}
$$

$$
F_r = F_r(\alpha_r) = F_r\left(\delta_r - \tan^{-1}\left(\tan \beta + \frac{l_r r}{V} \right) \right). \tag{2.59}
$$

The distance traveled in the OXY coordinate frame is calculated using

$$
\Delta X = \int_0^{t_f} V \cos(\beta + \Psi)dt, \tag{2.60}
$$

$$
\Delta Y = \int_0^{t_f} V \sin(\beta + \Psi)dt, \tag{2.61}
$$

where t_f is the final time corresponding to the distance ΔX and ΔY traveled. Due to the absence of longitudinal tire forces (and a powertrain model to drive it), the nonlinear single-track model will have deceleration (decreasing speed) when the steering angle is not zero. To avoid this situation in simulations and to be able to compare simulation results with the linearized single-track model where vehicle speed is constant, a simple driver model that regulates longitudinal speed is needed. The simplest cruise controller or driver model that can be used is

$$
F_x \rightarrow F_x + G_{cruise}(s)(V_{des} - V), \tag{2.62}
$$

where V_{des} is the desired speed and $G_{cruise}(s)$ is a longitudinal speed controller which can be a proportional-derivative (PD) or a high gain proportional (P) controller, for example.

2.3.3 Augmented Single-Track Lateral Vehicle Model

The nonlinear single-track vehicle model presented above can be augmented by the addition of the longitudinal road load and longitudinal tire forces to obtain a higher fidelity model to be used in simulation studies. The augmented single-track vehicle model is illustrated in Figure 2.24. The right-hand sides of the ΣF_x, ΣF_y, and ΣM_z equations ((2.46) and (2.47)) change as

$$\sum F_x = -F_{yf}(s_f, \alpha_f) \sin \delta_f - F_{yr}(s_r, \alpha_r) \sin \delta_r + F_{xf}(s_f, \alpha_f) \cos \delta_f + F_{xr}(s_r, \alpha_r) \cos \delta_r$$
$$- \frac{1}{2}\rho_a C_d A_f(v_x + v_{wind})^2 - C_r(v_x)Mg \cos \theta - Mg \sin \theta,$$

$$\sum F_y = F_{yf}(s_f, \alpha_f) \cos \delta_f + F_{yr}(s_r, \alpha_r) \cos \delta_r + F_{xf}(s_f, \alpha_f) \sin \delta_f + F_{xr}(s_r, \alpha_r) \sin \delta_r, \tag{2.63}$$

and

$$\sum M_z = F_{yf}(s_f, \alpha_f)l_f \cos \delta_f - F_{yr}(s_r, \alpha_r)l_r \cos \delta_r$$
$$+ F_{xf}(s_f, \alpha_f)l_f \sin \delta_f - F_{xr}(s_r, \alpha_r)l_r \sin \delta_r + M_{zd}, \tag{2.64}$$

which can be expressed more compactly as

$$\begin{bmatrix} \sum F_x \\ \sum F_y \\ \sum M_z \end{bmatrix} = \underbrace{\begin{bmatrix} -\sin \delta_f & -\sin \delta_r & \cos \delta_f & \cos \delta_r \\ \cos \delta_f & \cos \delta_r & \sin \delta_f & \sin \delta_r \\ l_f \cos \delta_f & -l_r \cos \delta_r & l_f \sin \delta_f & -l_f \sin \delta_r \end{bmatrix}}_{\mathbf{T_{asp}}(\delta_f, \delta_r)} \begin{bmatrix} F_{yf} \\ F_{yr} \\ F_{xf} \\ F_{xr} \end{bmatrix} + \begin{bmatrix} -F_{RL} \\ 0 \\ 0 \end{bmatrix} + \begin{bmatrix} 0 \\ 0 \\ M_{zd} \end{bmatrix}, \tag{2.65}$$

where F_{RL} is the sum of the longitudinal road load terms, and $\mathbf{T_{asp}}(\delta_f, \delta_r)$ is the augmented steering projection similar to the steering projection in Eq. (2.48). Since the tire forces are now functions of both longitudinal slip and sideslip, the longitudinal slip values for the front and rear tires also need to be calculated as

$$s_f = \frac{R\omega_f - v_{fxt}}{\max(v_{fxt}, R\omega_f, \varepsilon)}, \tag{2.66}$$

$$s_r = \frac{R\omega_r - v_{rxt}}{\max(v_{rxt}, R\omega_r, \varepsilon)}, \tag{2.67}$$

$$v_{fxt} = v_f \cos \alpha_f, \quad v_f = |\mathbf{v_f}| = \sqrt{(V \cos \beta)^2 + (V \sin \beta + rl_f)^2}, \tag{2.68}$$

$$v_{rxt} = v_r \cos \alpha_r, \quad v_f = |\mathbf{v_r}| = \sqrt{(V \cos \beta)^2 + (V \sin \beta - rl_r)^2}. \tag{2.69}$$

The new kinematics/geometry block will also calculate the longitudinal tire slips using Eqs. (2.66) and (2.67). These computations, in turn, require the computation of tire rotational speeds which need front- and rear-wheel moment equations and drive and brake torque application, i.e. simple powertrain and braking models like the ones used in longitudinal dynamics modeling.

2.3.4 Linearized Single Track Lateral Vehicle Model

Linear plant models are most useful for controller design. Our approach will be to design controllers for linearized or linear speed-varying models and test using higher fidelity simulation models and

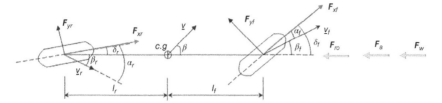

Figure 2.24 Augmented single-track vehicle model.

make sure that neglected model uncertainty is accommodated for by the robustness of the designed controller. For this purpose, the nonlinear single-track model presented in Section 2.3.2 will be linearized in this section. All angles will be linearized using small angle assumptions in radians. However, vehicle center of mass speed V will still appear in a nonlinear fashion in Eqs. (2.56), (2.58), and (2.59) as a $1/V$ term in several places. The only way to obtain a linear model will be to assume that vehicle speed V is constant. By keeping V as a variable parameter in the linearized single-track vehicle lateral dynamics model, a parameter (speed) varying linear model will be obtained. The linear model will have two state variables β and r instead of three in its nonlinear counterpart. After the small angle linearizations, the state space representation of the linearized single-track model is obtained as

$$\begin{bmatrix} \dot{\beta} \\ \dot{r} \end{bmatrix} = \begin{bmatrix} \dfrac{-C_f - C_r}{MV} & -1 + \left(\dfrac{C_r l_r - C_f l_f}{MV^2} \right) \\ \dfrac{C_r l_r - C_f l_f}{I_z} & \dfrac{-C_f l_f^2 - C_r l_r^2}{I_z V} \end{bmatrix} \begin{bmatrix} \beta \\ r \end{bmatrix} + \begin{bmatrix} \dfrac{C_f}{MV} & \dfrac{C_r}{MV} \\ \dfrac{C_f l_f}{I_z} & \dfrac{C_r l_r}{I_z} \end{bmatrix} \begin{bmatrix} \delta_f \\ \delta_r \end{bmatrix} + \begin{bmatrix} 0 \\ \dfrac{1}{I_z} \end{bmatrix} M_{zd}. \tag{2.70}$$

The transfer function from the FWS angle δ_f to the yaw rate r is obtained as

$$\frac{r(s)}{\delta_f(s)} = G(s, V) = \frac{b_1(V)s + b_0(V)}{a_2(V)s^2 + a_1(V)s + a_0(V)}, \tag{2.71}$$

where the coefficients are given by

$$b_0(V) = C_f C_r(l_f + l_r)V, \quad b_1(V) = C_f l_f MV^2$$
$$a_0(V) = C_f C_r(l_f + l_r)^2 + (C_r l_r - C_f l_f)MV^2$$
$$a_1(V) = \left(C_f \left(I_z + l_f^2 M \right) + c_r \left(I_z + l_r^2 M \right) \right) V, \quad a_2(V) = I_z MV^2. \tag{2.72}$$

The yaw disturbance input transfer function from yaw disturbance moment M_{zd} to yaw rate r for the linearized single-track model is

$$\frac{r(s)}{M_{zd}(s)} = G_d(s, v) = \frac{mV^2 s + (C_f + C_r)V}{a_2 s^2 + a_1 s + a_0}. \tag{2.73}$$

Note that the transfer function from a corrective yaw moment that can be generated by steering or individual braking intervention to yaw rate is the same as Eq. (2.73).

Handling properties of road vehicles are evaluated using the steady-state characteristics of the linearized single-track vehicle lateral model. Using Eq. (2.71), the steady-state characteristic is obtained by setting $s = 0$ as

$$\frac{r(s)}{\delta_f(s)} \bigg|_{s=0} = \frac{b_1(V)s + b_0(V)}{a_2(V)s^2 + a_1(V)s + a_0(V)} \bigg|_{s=0} = \frac{b_0(V)}{a_0(V)}, \tag{2.74}$$

which can also be expressed using the definitions (2.72) as

$$\left(\frac{r}{\delta_f}\right)_{ss} = \lim_{s\to 0} G(s,V)\bigg|_{\mu=1} \equiv K_n(V) = \frac{b_0(V)}{a_0(V)}\bigg|_{\mu=1} = \frac{C_f C_r(l_f+l_r)V}{C_f C_r(l_f+l_r)^2 + (C_r l_r - C_f l_f)MV^2}\bigg|_{\mu=1},$$

(2.75)

where ss denotes steady state and $\mu = 1$ is specified to denote dry road conditions. Equation (2.75), thus, defines the steady-state yaw rate gain of a vehicle under dry road conditions and is traditionally used to calculate the desired yaw rate in response to a steering wheel input. This is the steering response that a driver expects from the vehicle under nonextreme driving conditions. The linear single-track model and its steady-state gain are good approximations of lateral vehicle dynamics up to 0.3g to 0.4g of lateral acceleration where g is the gravitational acceleration.

Equation (2.75) can be re-expressed as

$$\left(\frac{r}{\delta_f}\right)_{ss} = K_n(V) = \frac{V}{L + K_{us}\frac{V^2}{g}},$$

(2.76)

where

$$K_{us} \equiv \left(\frac{l_r}{C_f L} - \frac{l_f}{C_r L}\right) Mg$$

(2.77)

is the understeer gradient. A negative understeer gradient vehicle with $K_{us} < 0$ has oversteer characteristic. A positive understeer gradient $K_{us} > 0$ has understeer characteristic. When $K_{us} = 0$, the vehicle has neutral steer characteristics with

$$\frac{l_r}{C_f L} - \frac{l_f}{C_r L} = 0$$

(2.78)

and

$$\frac{r}{\delta_f} = \frac{V}{L}, \quad r = \frac{V}{R_p}, \quad \delta_f = \frac{L}{R_p},$$

(2.79)

which is the same as Eq. (2.43). Since the lateral acceleration is $a_y = V^2/R_p$ for a circular path and since yaw rate is $r = V/R_p$, $r = a_y/V$ and Eq. (2.76) can be expressed as

$$\left(\frac{a_y}{\delta_f}\right)_{ss} = \frac{V^2}{L + K_{us}\frac{V^2}{g}}.$$

(2.80)

Solving Eq. (2.76) for δ_f and noting the second identity in Eq. (2.79), we obtain

$$\delta_f = \frac{L}{R_p} + K_{us}\frac{V^2}{R_p g} = \frac{L}{R_p} + K_{us}\frac{a_y}{g}.$$

(2.81)

The yaw gain in Eq. (2.76) is plotted against vehicle speed for an oversteer and understeer vehicle. For an oversteer vehicle, speed can only be increased up to the critical speed at which point the yaw gain blows up. This means that as we increase speed in an oversteer vehicle during cornering, the vehicle will start spinning as we reach the critical speed V_{cr} which is obtained by setting the denominator expression in Eq. (2.76) equal to zero to obtain

$$L + K_{us}(V_{cr})^2/g = 0, \quad V_{cr} = \sqrt{-Lg/K_{us}}.$$

(2.82)

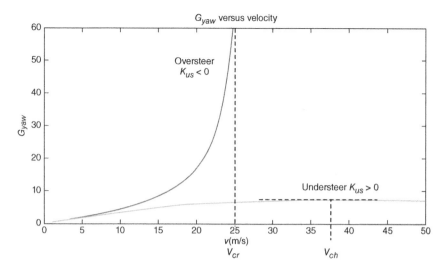

Figure 2.25 Yaw gain versus speed for oversteer and understeer vehicles.

The critical speed is only defined for an oversteer vehicle as K_{us} needs to be negative in Eq. (2.82). The yaw gain of an understeer vehicle has a maximum at the characteristic speed V_{ch} as shown in Figure 2.25. Setting the derivative of the yaw gain in Eq. (2.76) equal to zero results in

$$\frac{d}{dt}\left(\frac{V}{L + K_{us}V^2/g}\right) = 0 \Rightarrow L - K_{us}(V_{ch})^2/g = 0, \quad V_{ch} = \sqrt{Lg/K_{us}}. \tag{2.83}$$

The characteristic speed makes sense only for an understeer vehicle with $K_{us} > 0$. As vehicle speed is increased beyond the characteristic speed, yaw gain decreases, meaning that steering is less responsive as compared to lower speeds and a higher steering angle is needed for the same steady-state yaw rate value. Note that the understeer gradient in Eq. (2.77) will change with changes in the center of mass location due to changes in loading and with changes in the cornering stiffness of the tires. This will make it harder for the autonomous vehicle path tracking controller which must have inherent robustness against vehicle model parameter changes. The nonlinear yaw gain changes with vehicle speed as shown in Figure 2.25 will also make that task of the path following controller more difficult and are best resolved by using a speed-scheduled controller.

2.4 Path Model

The desired path to be followed, the single-track vehicle model and the path distance and orientation error are illustrated in Figure 2.26. h is the deviation from the path measured at center of gravity of vehicle. $\Delta\psi_p$ is angle of the tangent to the path in radians. Our aim is to make both the distance offset h from the path and the orientation error $\Delta\psi_p$ go to zero. This is a single input and multiple output problem which is difficult and inconvenient to handle. Instead, let us define l_s as the speed-dependent preview distance and try to reduce the lateral deviation from the tangent to the path at the preview distance, e_y, go to zero. This is much more convenient and easier to handle as this is now a single-input, single-output control design problem. This also makes more sense as it is also how drivers steer their vehicle to follow a path. They use a speed-dependent gaze to estimate and correct the error e_y at the preview distance determined by their gaze. This is also a more stable approach as it adds phase advance to the path-tracking loop since future error is used.

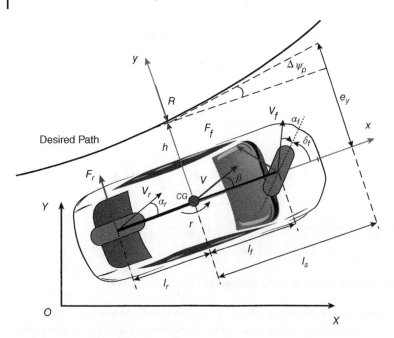

Figure 2.26 Desired path and tracking error.

The path-tracking error e_y at the preview distance in Figure 2.26 is given by

$$e_y = h + l_s(V) \tan(\Delta \psi_p). \tag{2.84}$$

The orientation error $\Delta \psi_p$ is zero for straight road and is a small value in radians if preview length l_s is kept small even for a large curvature. Keeping preview distance l_s small means that speed also has to be kept small. Equation (2.84) is linearized for small $\Delta \psi_p$ in radians as

$$e_y \cong h + l_s \Delta \psi_p. \tag{2.85}$$

We will add two more states to the single-track model as $\Delta \psi_p$ and e_y, so we will need to differentiate both of them. Derivatives of h and $\Delta \psi_p$ will be derived next as they are needed in obtaining the derivative of e_y.

The more detailed geometry of the single-track vehicle path tracking is shown in Figure 2.27. Four-wheel steering has been assumed in Figure 2.27. According to Figure 2.27, dh/dt is the speed component along the perpendicular direction to the tangent to the path and is given by

$$\frac{dh}{dt} = V \sin (\beta + \Delta \psi_p), \tag{2.86}$$

and is also the derivative of the path deviation at the center of gravity. This relation is linearized assuming small angles β and $\Delta \psi_p$ in radians resulting in

$$\frac{dh}{dt} \cong V (\beta + \Delta \psi_p). \tag{2.87}$$

The desired path has been translated along the normal direction to coincide with the vehicle center of mass such that the distance error h is zero in Figure 2.28. The inertial frame of reference OXY has also been translated to the center of mass of the vehicle such that O is now at CG in Figure 2.28. The component of vehicle velocity tangent to the path is seen to be $V \cos(\beta + \Delta \psi_p)$. The angle that the vehicle makes with the inertial X axis is $\psi - \Delta \psi_p$. Using Figure 2.28, the rate

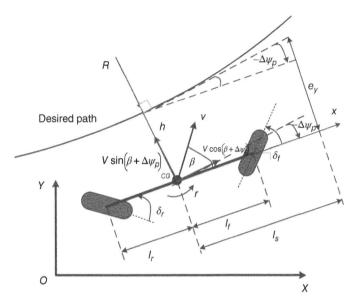

Figure 2.27 Path-tracking model detailed geometry.

of change of tangent to the desired path with time is seen to be the component of vehicle speed in direction of desired path tangent divided by the instantaneous radius of curvature of the path and is given by

$$\frac{d}{dt}\left(\psi - \Delta\psi_p\right) = \frac{V\,\cos\left(\beta + \Delta\psi_p\right)}{R}. \tag{2.88}$$

This can be re-expressed as

$$\frac{d\Delta\psi_p}{dt} = \frac{d\psi}{dt} - \frac{V\,\cos\left(\beta + \Delta\psi_p\right)}{R} = r - \frac{V\,\cos\left(\beta + \Delta\psi_p\right)}{R}, \tag{2.89}$$

noting that $d\psi/dt$ is the vehicle yaw rate r. Using the definition of curvature ρ_{ref} as being $1/R$, Eq. (2.89) becomes

$$\frac{d\Delta\psi_p}{dt} = r - V\,\cos\left(\beta + \Delta\psi_p\right)\rho_{ref}. \tag{2.90}$$

Linearizing for small angles β and $\Delta\psi_p$ in radians results in

$$\frac{d\Delta\psi_p}{dt} \cong r - V\rho_{ref}. \tag{2.91}$$

Differentiation of the path error e_y at the preview distance l_s given in Eq. (2.84) results in

$$\frac{de_y}{dt} = \frac{dh}{dt} + \frac{dl_s}{dV}\frac{dV}{dt}\tan\left(\Delta\psi_p\right) + l_s\left(V\right)\frac{d\left(\tan\left(\Delta\psi_p\right)\right)}{d\left(\Delta\psi_p\right)}\frac{d\left(\Delta\psi_p\right)}{dt}. \tag{2.92}$$

Using Eq. (2.86) for dh/dt and Eq. (2.90) for $d\Delta\psi_p/dt$ in Eq. (2.92), we obtain

$$\begin{aligned}
\frac{de_y}{dt} &= V\,\sin\left(\beta + \Delta\psi_p\right) + \frac{dl_s}{dV}\frac{dV}{dt}\tan\left(\Delta\psi_p\right) \\
&\quad + \frac{l_s(V)}{\cos^2\left(\Delta\psi_p\right)}\left(r - \left(V\,\cos\left(\beta + \Delta\psi_p\right)\rho_{ref}\right)\right).
\end{aligned} \tag{2.93}$$

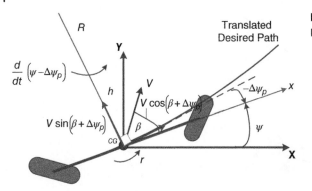

Figure 2.28 Desired path translated to pass through vehicle center of mass.

Equations (2.93) and (2.90) are the equations for the derivatives of the two new state variables e_y and $\Delta\psi_p$ which along with Eq. (2.57) for the other three state variable β, V, and r form the nonlinear path model based on the single-track vehicle lateral model. Linearization of Eq. (2.93) for small angles in radians and constant vehicle speed V results in the linear version

$$\frac{de_y}{dt} \cong V\beta + l_s r + V\Delta\psi_p - l_s V\rho_{ref}. \tag{2.94}$$

The linear single-track vehicle model (2.70) and the linearized equations (2.91) and (2.94) result in the linear path tracking state space model

$$
\begin{bmatrix} d\beta/dt \\ dr/dt \\ d\Delta\psi_p/dt \\ de_y/dt \end{bmatrix} = \underbrace{\begin{bmatrix} a_{11} & a_{12} & 0 & 0 \\ a_{21} & a_{22} & 0 & 0 \\ 0 & 1 & 0 & 0 \\ V & l_s(V) & V & 0 \end{bmatrix}}_{\text{vehicle and path dynamics}} \begin{bmatrix} \beta \\ r \\ \Delta\psi_p \\ e_y \end{bmatrix} + \underbrace{\begin{bmatrix} b_{11} & b_{12} \\ b_{21} & b_{22} \\ 0 & 0 \\ 0 & 0 \end{bmatrix} \begin{bmatrix} \delta_f \\ \delta_r \end{bmatrix}}_{\text{steering command}}
$$

$$
+ \underbrace{\begin{bmatrix} 0 \\ 0 \\ -V \\ -l_s(V)V \end{bmatrix} \rho_{ref}}_{\text{road curvature disturbance}} + \underbrace{\begin{bmatrix} 0 \\ \frac{1}{I_z} \\ 0 \\ 0 \end{bmatrix} M_{zd}}_{\text{yaw moment disturbance}}, \tag{2.95}
$$

where

$$a_{11} = \frac{-C_f - C_r}{MV}, \quad a_{12} = -1 + \left(\frac{C_r l_r - C_f l_f}{MV^2} \right), \quad b_{11} = \frac{C_f}{MV}, \quad b_{12} = \frac{C_r}{MV},$$

$$a_{21} = \frac{C_r l_r - C_f l_f}{I_z}, \quad a_{22} = \frac{-C_f l_f^2 - C_r l_r^2}{I_z V}, \quad b_{21} = \frac{C_f l_f}{I_z}, \quad b_{22} = \frac{C_r l_r}{I_z}. \tag{2.96}$$

For a front-wheel steering vehicle, δ_r is zero and path-tracking controller design uses the single-input, single-output loop from δ_f to e_y which is chosen as the output in the presence of path curvature disturbance ρ_{ref} and yaw moment disturbance M_{zd}. Parametric transfer functions like the ones in Eqs. (2.71) and (2.73) are very useful for controller design purposes, especially for speed scheduled robust controller design. Parametric transfer functions corresponding to Eq. (2.95) are fourth order and quite long due to the number of terms involved. They can be expressed in compact notation as

$$e_y = G_{y\delta}(s, V)\delta_f + G_{y\rho}(s, V)\rho_{ref} + G_{yM}(s, V)M_{zd}, \qquad (2.97)$$

where a well-designed path-tracking controller will make preview length projected path following e_y go to zero while rejecting the effect of disturbances ρ_{ref} and M_{zd}.

The computation of the path deviation error e_y at the preview distance is presented next. The minimum distance from the vehicle center of mass to the path is computed first and is illustrated in Figure 2.29. Dimensionless parameter λ is used to characterize different points on the desired path such that any point P on the desired path has position vector and coordinates $\mathbf{R}_P(\lambda) = (X_P(\lambda), Y_P(\lambda))$. The desired path $\mathbf{R}_P(\lambda)$ with $\lambda_{min} < \lambda < \lambda_{max}$ is obtained from a map or a curve fit of experimentally collected way points. $\mathbf{R}_{CG} = (X_{CG}, Y_{CG})$ denotes the position vector of the center of mass of the vehicle and is obtained at each instant of time using a localization sensor like GPS or lidar or visual Simultaneous Localization and Mapping (SLAM) or scan matching based on a point cloud map. The orientation ψ of the vehicle is also determined by using successive GPS readings, differential GPS, or SLAM or map matching. The triplet (X_{CG}, Y_{CG}, ψ) is the pose of the autonomous vehicle at that instance. Based on Figure 2.29, the minimum distance on the path vector $\mathbf{R}_P - \mathbf{R}_{CG}$ is perpendicular to the derivative of $\mathbf{R}_P$ which is tangent to the path, expressed as

$$(\mathbf{R}_P(\lambda) - \mathbf{R}_{CG}) \cdot \dot{\mathbf{R}}_P = 0 \qquad (2.98)$$

at each instant of time. Equation (2.98) is manipulated further as

$$\begin{pmatrix} X_P(\lambda) - X_{CG} \\ Y_P(\lambda) - Y_{CG} \end{pmatrix} \cdot \begin{pmatrix} \dot{X}_P(\lambda) \\ \dot{Y}_P(\lambda) \end{pmatrix} = (X_P(\lambda) - X_{CG})\dot{X}_P(\lambda) + (Y_P(\lambda) - Y_{CG})\dot{Y}_P(\lambda) = 0, \qquad (2.99)$$

which has to be solved at each instant of time, i.e. for the current value of vehicle position given by X_{CG}, Y_{CG} to obtain the corresponding value λ. λ is then used in

$$h = |\mathbf{R}_P(\lambda) - \mathbf{R}_{CG}| = \begin{vmatrix} X_P(\lambda) - X_{CG} \\ Y_P(\lambda) - Y_{CG} \end{vmatrix} = \sqrt{(X_P(\lambda) - X_{CG})^2 + (Y_P(\lambda) - Y_{CG})^2} \qquad (2.100)$$

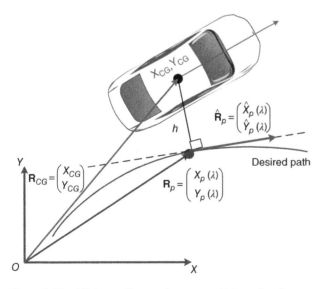

Figure 2.29 Minimum distance between vehicle and path.

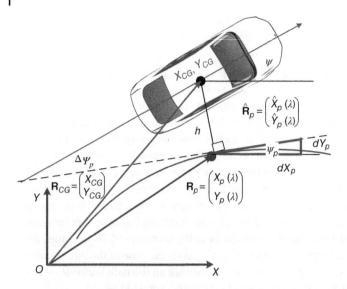

Figure 2.30 Angular orientation of path.

which is the distance h at the corresponding time instance. This computation has to be repeated at each step of the solution, simulation, or real time implementation.

The path angular orientation for the λ value determined in the minimum distance computation presented above will be determined next. Referring to Figure 2.30, the orientation angle of the path ψ_p is

$$\tan \psi_p = \frac{dY_P(\lambda)}{dX_P(\lambda)} = \frac{dY_P(\lambda)/dt}{dX_P(\lambda)/dt} = \frac{\dot{Y}_P(\lambda)}{\dot{X}_P(\lambda)}. \tag{2.101}$$

ψ_p is approximated by

$$\psi_p \cong \frac{\dot{Y}_P(\lambda)}{\dot{X}_P(\lambda)} \tag{2.102}$$

for small values in radians. Note that the path-tracking orientation error is

$$\Delta\psi_p = \psi - \psi_p, \tag{2.103}$$

which becomes

$$\Delta\psi_p \cong \psi - \frac{\dot{Y}_P(\lambda)}{\dot{X}_P(\lambda)}, \tag{2.104}$$

when (2.102) is used for ψ_p and is the equation used to calculate the path orientation error at each instance after λ is determined for minimum distance to path. Substitution of $\Delta\psi_p$ in the linearized path following error equation (2.85) results in

$$e_y \cong h + l_s \left(\psi - \frac{\dot{Y}_P(\lambda)}{\dot{X}_P(\lambda)} \right) \tag{2.105}$$

for the error at that instance corresponding to the computed value of λ. In summary, at each instance of time, Eq. (2.99) needs to be solved for λ, which will be used in Eq. (2.100) to determine h and in Eq. (2.104) to determine $\Delta\psi_p$, which are in turn used in Eq. (2.105) to determine the path-tracking error.

Note that desired path curvature appears as a disturbance input in the path following model of Eq. (2.95). The road curvature ρ_{ref} for a path parametrized by $\mathbf{R}_P(\lambda) = (X_P(\lambda), Y_P(\lambda))$ with $\lambda_{min} < \lambda < \lambda_{max}$ is calculated using

$$\rho_{ref}(\lambda) = \frac{\left| \dfrac{dX_P(\lambda)}{d\lambda} \dfrac{d^2 Y_P(\lambda)}{d\lambda^2} - \dfrac{dY_P(\lambda)}{d\lambda} \dfrac{d^2 X_P(\lambda)}{d\lambda^2} \right|}{\left(\left(\dfrac{dX_P(\lambda)}{d\lambda} \right)^2 + \left(\dfrac{dY_P(\lambda)}{d\lambda} \right)^2 \right)^{\frac{3}{2}}}. \tag{2.106}$$

The desired path model discussed in this section is two-dimensional, meaning that it is on a horizontal plane and elevation changes in the vertical direction were neglected to simplify the analysis and the exposition of results. It is straightforward to extend the formulas and approach used here to three-dimensional desired path models and path following error computation.

2.5 Pure Pursuit: Geometry-Based Low-Speed Path Tracking

The desired path following minimum distance to path, preview distance, and orientation error computations presented in Section 2.4 require some computation and the solution of Eq. (2.98) is not trivial. Feedback controllers used for path tracking require those computations as they use the preview distance path error e_y at each time step of simulation or real-time calculation. There are some very simple feedback control laws based on the geometry of turning and basic kinematics that do not need those computations. One such method, the pure pursuit, will be presented in this section as it is very easy to implement and works well at low speeds where high accuracy is not necessary. It is also very useful for educational purposes and for forming a fast benchmark solution for comparison purposes.

Consider the left subplot in Figure 2.31 first. Suppose that we want our autonomous vehicle to go from its initial location to the final location shown with a large dot. Denote the initial position of the vehicle by A and the desired location by B as in the middle subplot in Figure 2.31 and draw the circular arc connecting A and B assuming circular turning for a fixed steering wheel angle. Denote the secant length as l_s which will be equivalent to our path preview distance. Draw the rest of the circular arc to obtain a complete circle as shown in the right subplot in Figure 2.31 and note that

$$X_P^2 + Y_P^2 = l_s^2, \quad X_P + d = R, \quad d = R - X_P \tag{2.107}$$

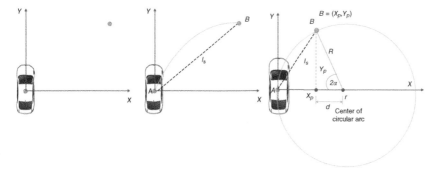

Figure 2.31 Pure pursuit steering.

for the right triangle in Figure 2.31. The second right triangle in Figure 2.31 has

$$d^2 + Y_P^2 = (R - X_P)^2 + Y_P^2 = R^2. \tag{2.108}$$

The middle part of Eq. (2.108) is expanded and simplified into

$$-2RX_P + X_P^2 + Y_P^2 = 0. \tag{2.109}$$

Using the first equality in Eq. (2.107) and solving for R results in

$$R = \frac{l_s^2}{2X_P}. \tag{2.110}$$

Using the definition of the radius of curvature in Eq. (2.110) results in

$$\rho_{ref} = \frac{1}{R} = \frac{2X_P}{l_s^2}. \tag{2.111}$$

Similar to Eq. (2.43) for Ackerman steering

$$\delta_f = \tan^{-1}\left(\frac{L}{R}\right) \cong \frac{L}{R}, \tag{2.112}$$

and using Eq. (2.111) for $1/R$ results in the pure pursuit steering control law

$$\delta_f = \tan^{-1}\left(\frac{2LX_P}{l_s^2}\right) \cong \frac{2L}{l_s^2}X_P \equiv K_P X_P, \tag{2.113}$$

which is seen to be a proportional controller with gain

$$K_P \equiv \frac{2L}{l_s^2} = \frac{2L}{(l_s(X_P, Y_P))^2}. \tag{2.114}$$

The simple pursuit control law in Eq. (2.113) is a point-to-point-motion controller. The pair (X_P, Y_P) is always the next waypoint to be followed so that at each instant in time the steering angle is computed to drive the vehicle toward the next waypoint. This results in a path-tracking motion with sudden stepwise changes in steering and accompanying nonsmooth motion. The computations and implementation are very simple, however, and pure pursuit is a useful method for low-speed mobile robotics tasks. The commonly used and known equivalent form of the pure pursuit path-tracking control law is obtained from Eq. (2.113) by using the angle α in Figure 2.31 and is given by

$$\delta_f = \tan^{-1}\left(\frac{2L}{l_s}\sin \alpha\right). \tag{2.115}$$

2.6 Stanley Method for Path Tracking

The second path-tracking method that will be presented in this chapter is the Stanley method which was used by the Stanford Darpa Grand Challenge Team in 2005 [6]. This is a more advanced method as compared to the simple point-to-point pure pursuit tracking and is similar in nature to the path error computation presented in Section 2.4 and is illustrated in Figure 2.32. Note that this is very similar to Figure 2.26. The main differences are that the path distance error is now calculated at the front axle instead of the center of mass so that h_f is calculated and used as shown in Figure 2.32 as compared to h in Figure 2.26. The computation for h_f is similar to the previous computation as it necessitates finding the minimum distance from the front axle of the single-track vehicle model to the path. h_f is still on the direction that is perpendicular to the local tangent to the path at the point that is closest in distance to the front axle. The computation methods and formulas in Section 2.4

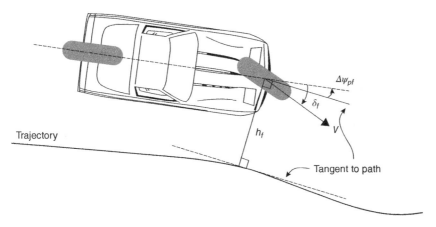

Trajectory

$\Delta\psi_{pf}$

δ_f

h_f

V

Tangent to path

Figure 2.32 Stanley method.

can be used after replacing the vehicle center of mass coordinates with the front axle coordinates. The angular orientation error is denoted as $\Delta\psi_{pf}$ now as it is the orientation mismatch between the vehicle orientation and the path tangent at the minimum distance from the front axle. So the computations have been moved from the vehicle center of mass to the front axle. The other difference as compared to Section 2.4 is that a preview distance is not used anymore. The geometry and kinematics based part of the Stanley path tracking control law is

$$\delta_f = \Delta\psi_{pf} + \tan^{-1}\left(K\frac{h_f}{V_\varepsilon + V}\right) \cong \underbrace{\Delta\psi_{pf}}_{\substack{\text{compensates} \\ \text{orientation error}}} + \underbrace{K_p(V)h_f}_{\substack{\text{like speed scheduled} \\ \text{pure pursuit for offset} \\ \text{distance error}}},$$

(2.116)

with the last expression being the approximation for small argument values in the inverse tangent function and where $K_p(V) = K/(V_\varepsilon + V)$. The steering is adjusted to be the same as the path orientation error using the first term $\Delta\psi_{pf}$ in Eq. (2.115). The second term $K_p h_f$ in Eq. (2.115) is a proportional control term to reduce distance offset error h_f to zero like the pure pursuit control law. Unlike the pure pursuit control law in Eq. (2.112), the proportional gain is speed scheduled and decreases with speed as it should. V_ε is used to make sure that the Stanley method steering angle does not blow up at low speeds. It is also used to change the proportional gain to a preselected constant value at low speeds. A value of $V_\varepsilon = 1$ m/s was used in the original Stanley steering controller in the 2005 Darpa Urban Challenge. The original 2005 Darpa Urban Challenge Stanley steering formula also had other dynamic feedback components as

$$\delta_f = \underbrace{\Delta\psi_{pf} + K_p(V)h_f}_{\substack{\text{commonly used geometric} \\ \text{kinematic Stanley formula}}} + \underbrace{K_{dr}(r - r_{dp})}_{\substack{\text{yaw dynamics} \\ \text{improvement}}} + \underbrace{K_s(\delta_f(i) - \delta_f(i+1))}_{\text{adds positive phase to steering}},$$

(2.117)

where r_{dp} is the yaw rate of the desired path being followed. The yaw dynamics improvement is needed since the geometric/kinematic part of the Stanley controller (first two terms in Eq. (2.116)) will not be able to do a good job of yaw regulation as computations are based on distance and orientation error at the front axle. The path-tracking formulation in Section 2.4 does not have this problem as computations are with respect to the vehicle center of mass. The Stanley controller without the yaw compensation term will work well for short wheelbase vehicles at low speeds for paths of low curvature. Large wheelbase, higher speeds, and large curvature paths will necessitate the use of the yaw compensation term. The last expression in Eq. (2.117) is used to improve steering dynamics. Rearranging followed by taking the Laplace transform of both sides in Eq. (2.117) assuming constant speed, using the same notation for time domain and Laplace domain variables and rearranging again results in

$$\delta_f = \Delta\psi_{pf} + K_p(V)h_f + K_{dr}(r - r_{dp}) - \underbrace{(K_s(t_{i+1} - t_i))}_{\tau_s}\left(\frac{\delta_f(i+1) - \delta_f(i)}{t_{i+1} - t_i}\right), \tag{2.118}$$

$$\delta_f = \Delta\psi_{pf} + K_p(V)h_f + K_{dr}(r - r_{dp}) - \tau_s\frac{d\delta_f}{dt}, \tag{2.119}$$

$$\delta_f = \frac{1}{\tau_s + 1}\left(\Delta\psi_{pf} + K_p(V)h_f + K_{dr}(r - r_{dp})\right), \tag{2.120}$$

which shows that the steering improvement in Eq. (2.117) is nothing but the application of a unity d.c. gain first-order low pass filter with cutoff frequency $1/\tau_s$ to the geometry/kinematics plus yaw compensation Stanley steering control law.

The Stanley formula path distance offset at the front axle can be calculated in a similar manner to Section 2.4 by replacing the center of mass location with the front axle in the computation. The equations become

$$(\mathbf{R}_P(\lambda) - \mathbf{R}_F) \cdot \dot{\mathbf{R}}_P = 0, \tag{2.121}$$

where $R_F = (X_F, Y_F)$ is the location of the center of the front axle or the front tire location in the single-track vehicle model. Equation (2.121) expresses the fact that $\mathbf{R}_P - \mathbf{R}_F$ is perpendicular to the derivative of $\mathbf{R}_P$ which is tangent to the path. Equation (2.121) is manipulated further as

$$\begin{pmatrix} X_P(\lambda) - X_F \\ Y_P(\lambda) - Y_F \end{pmatrix} \cdot \begin{pmatrix} \dot{X}_P(\lambda) \\ \dot{Y}_P(\lambda) \end{pmatrix} = (X_P(\lambda) - X_F)\dot{X}_P(\lambda) + (Y_P(\lambda) - Y_F)\dot{Y}_P(\lambda) = 0, \tag{2.122}$$

which has to be solved at each instant of time, i.e. for the current value of vehicle position given by X_F, Y_F to obtain the corresponding value λ. λ is then used in

$$h_f = |\mathbf{R}_P(\lambda) - \mathbf{R}_F| = \left|\begin{matrix} X_P(\lambda) - X_F \\ Y_P(\lambda) - Y_F \end{matrix}\right| = \sqrt{(X_P(\lambda) - X_F)^2 + (Y_P(\lambda) - Y_F)^2}, \tag{2.123}$$

to obtain the minimum path offset distance h_f between the front axle and the path at the corresponding time instance. This computation has to be repeated at each step of the solution, simulation, or real-time implementation. Similar to Section 2.4, the path orientation error at the front axle is calculated using

$$\Delta\psi_{pf} \cong \psi - \frac{\dot{Y}_P(\lambda)}{\dot{X}_P(\lambda)}. \tag{2.124}$$

2.7 Path Tracking in Reverse Driving and Parking

The modeling used in Section 2.4 and Eq. (2.95) can be modified easily to formulate driving backwards in reverse which can also be used for reverse maneuvers by switching δ_f and δ_r in those equations. This is illustrated in Figure 2.33. This is like a rear-wheel steered vehicle now, l_f and

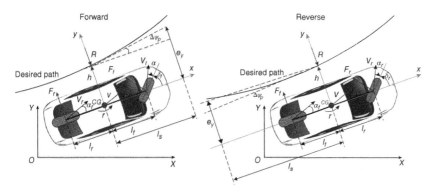

Figure 2.33 Path tracking in reverse driving.

l_r and C_f and C_r are also interchanged and the preview distance l_s is now toward the rear of the vehicle. The path-tracking model becomes

$$
\begin{bmatrix} d\beta/dt \\ dr/dt \\ d\Delta\psi_p/dt \\ de_y/dt \end{bmatrix} = \underbrace{\begin{bmatrix} a_{11r} & a_{12r} & 0 & 0 \\ a_{21r} & a_{22r} & 0 & 0 \\ 0 & 1 & 0 & 0 \\ V & l_s(V) & V & 0 \end{bmatrix}}_{\substack{\text{vehicle and path dynamics} \\ \text{for reverse driving}}} \begin{bmatrix} \beta \\ r \\ \Delta\psi_p \\ e_y \end{bmatrix} + \underbrace{\begin{bmatrix} b_{11r} & b_{12r} \\ b_{21r} & b_{22r} \\ 0 & 0 \\ 0 & 0 \end{bmatrix} \begin{bmatrix} \delta_r \\ \delta_f \end{bmatrix}}_{\text{reverse steering command}}
$$

$$
+ \underbrace{\begin{bmatrix} 0 \\ 0 \\ -V \\ -l_s(V)V \end{bmatrix} \rho_{ref}}_{\substack{\text{road curvature} \\ \text{disturbance}}} + \underbrace{\begin{bmatrix} 0 \\ \frac{1}{I_z} \\ 0 \\ 0 \end{bmatrix} M_{zd}}_{\substack{\text{yaw moment} \\ \text{disturbance}}} ,
\tag{2.125}
$$

where the terms inside the matrices are

$$
a_{11} = \frac{-C_f - C_r}{MV}, \quad a_{12} = -1 + \left(\frac{C_f l_f - C_r l_r}{MV^2} \right), \quad b_{11} = \frac{C_r}{MV}, \quad b_{12} = \frac{C_f}{MV},
$$

$$
a_{21} = \frac{C_f l_f - C_r l_r}{I_z}, \quad a_{22} = \frac{-C_r l_r{}^2 - C_f l_f{}^2}{I_z V}, \quad b_{21} = \frac{C_r l_r}{I_z}, \quad b_{22} = \frac{C_f l_f}{I_z}.
\tag{2.126}
$$

Note that the understeer gradient also changes as

$$
K_{us,reverse} \equiv \left(\frac{l_r}{C_f L} - \frac{l_f}{C_r L} \right) Mg,
\tag{2.127}
$$

which is the negative of the forward-driving understeer gradient in Eq. (2.77). This means that a typical understeer car in forward driving will become and oversteer car in reverse driving, explaining part of the reason for reverse driving being more difficult that forward driving. Reverse driving has to be confined to lower speeds, definitely less that the critical speed with the new understeer gradient in Eq. (2.127).

2.8   Chapter Summary and Concluding Remarks

Vehicle and path models were derived and presented in this chapter for use in path-tracking simulations, tracking controller design, and analysis in Chapters 5–7. The presented models include linear, linear parameter varying, and nonlinear formulations. The varying parameter is vehicle speed. The linear and linear parameter varying models will be useful for analysis and controller design while the nonlinear models will be useful for simulation validation. Combined slip tire force models with tire force saturation were also presented along with the friction/Kamm ellipse or circle concept. Validation of designed controllers using realistic combined slip tire force models is important in designing autonomous vehicle path-tracking controllers that will work satisfactorily during actual operation on roads.

References

1 E. Dincmen, B. Aksun-Guvenc, and T. Acarman, "Extremum seeking control of ABS braking in road vehicles with lateral force improvement," *IEEE Transactions on Control Systems Technology*, vol. 22, no. 1, pp. 230–237, 2014.

2 E. Bakker, L. Nyborg, and H. Pacejka, "Tyre modelling for use in vehicle dynamics studies," SAE Technical Paper 870421, 1987.

3 A. Guvenc, T. Acarman, and L. Guvenc, "Coordination of steering and individual wheel braking actuated vehicle yaw stability control," *IEEE IV2003 Intelligent Vehicles Symposium. Proceedings (Cat. No.03TH8683)*, 2003, pp. 288–293, https://doi.org/10.1109/IVS.2003.1212924.

4 R. Brach and R. Brach, "Modeling combined braking and steering tire forces," SAE Technical Paper 2000-01-0357, 2000.

5 B. Aksun-Guvenc, L. Guvenc, and S. Karaman, "Robust yaw stability controller design and hardware-in-the-loop testing for a road vehicle," *IEEE Transactions on Vehicular Technology*, vol. 58, no. 2, pp. 555–571, 2009.

6 G. M. Hoffmann, C. J. Tomlin, M. Montemerlo, and S. Thrun, "Autonomous automobile trajectory tracking for off-road driving: controller design, experimental validation and racing," *2007 American Control Conference*, 2007, pp. 2296–2301, https://doi.org/10.1109/ACC.2007 .4282788.

3

Simulation, Experimentation, and Estimation Overview

This chapter presents background information that will be useful in the understanding of Chapters 4–7. Different types of simulations that need to be conducted using the vehicle and path following models developed in Chapter 2 and higher fidelity models are treated first. The concepts of model-in-the-loop (MIL), hardware-in-the-loop (HIL), traffic-in-the-loop (TIL), infrastructure-in-the-loop (IIL), and vehicle-in-the-loop (VIL) simulation are defined and introduced. The higher fidelity CarSim or similar modeling used in such simulations is presented. The HIL and vehicles used in the experimental results presented in this book are explained. The unified hardware/software architecture used in these vehicles is explained briefly. Creation of virtual environments for realistic generation of autonomous vehicle (AV) perception sensor data is presented using the Unity Engine and LGSVL simulator and the Unreal Engine and CARLA simulator. The chapter ends with a section on estimation of unknown and hard to measure parameters using the estimation of effective tire radius and tire-road friction estimation as examples.

3.1 Introduction to the Simulation-Based Development and Evaluation Process

Developing, constructing, and evaluating hardware and software for complex systems such as autonomous vehicles is a challenging task. In this book, we concentrate on path planning and path-tracking control for autonomous driving. Since an autonomous vehicle needs to follow a speed profile while tracking that path, the problem becomes one of trajectory planning and trajectory tracking control. Validation of the designed path or trajectory and the tracking controllers for the longitudinal and lateral dynamics loops is necessary in an evaluation process where all possible operating conditions, uncertainties, and disturbances must be considered. This requires an extensive amount of evaluation testing along with possible code, modeling, and design revisions as needed. In order to minimize the cost of this evaluation and validation procedure, the designs are first tested in virtual environments where soft models of the test environments and systems are created and validated. Then, these soft models work together with the designed path/trajectory planning and tracking control software where a combination of many different conditions are tested to see if there is any problem while they work together. This is a purely software-based approach using models of the vehicle and the environment and is thus called Model-in-the-Loop or MIL simulation. In some cases, the path/trajectory planning and tracking control software is coded and used in the simulations exactly as the code that will be implemented in a real deployment. The evaluation method is then called Software-in-the-Loop or SIL testing. Either Simulink models of tire and vehicle dynamics as presented in Chapter 2 or higher fidelity

Autonomous Road Vehicle Path Planning and Tracking Control, First Edition.
Levent Güvenç, Bilin Aksun-Güvenç, Sheng Zhu, and Şükrü Yaren Gelbal.
© 2022 The Institute of Electrical and Electronics Engineers, Inc. Published 2022 by John Wiley & Sons, Inc.

models like those in CarSim, CarMaker, or similar programs are used to represent the vehicle in the simulations.

The next step after MIL or SIL evaluation is to apply Hardware-in-the-Loop or HIL simulation. The difference from MIL simulation is the presence of at least one hardware component in the loop. In the simplest implementation of HIL simulation, one of the autonomous driving computation computers is placed inside the simulation loop and is connected physically in such a manner that a real time operating and hence fast computer running the vehicle model provides the needed signals during the simulation [1]. This is very useful for checking the real-time computability of autonomous driving algorithms like online path planning and modification and path tracking control. Interface hardware is usually needed for a realistic interface between the real-time vehicle dynamics simulation computer and the autonomous driving electronic control unit or in-vehicle computer. In the simplest implementation, just a controller area network (CAN) bus connection may suffice as the interface. Actuators and sometimes sensors like camera and radar or a communication modem are also added to the loop in HIL simulation. Road-side units and traffic control computers are added to the HIL simulator to create Infrastructure-in-the-Loop or IIL simulation [2]. Realistic environment modeling based on game engine rendering like Unity or Unreal are used to do Environment-in-the-Loop or EIL simulations which can be used in both MIL/SIL and HIL settings. This realistic environment modeling allows the generation of realistic data for camera and lidar sensors that can be used to evaluate perception algorithms. Microscopic traffic simulators are cosimulated in real time along with the environment and vehicle dynamics models to test the autonomous driving software within realistic traffic situations. This is called Traffic-in-the-Loop or TIL simulation [3, 4]. Sometimes the stationary vehicle is added to the simulation loop, resulting in Vehicle-in-the-Loop or VIL simulation [5]. If the vehicle is placed on a chassis dynamometer during the simulation, it is called Dyno-in-the-Loop or DIL simulation. An example is shown in Figure 3.1.

The steering mechanism of the vehicle is sometimes disconnected such that it does not turn the steerable wheels/tires in the dyno. The steering actuator still works and if it is connected

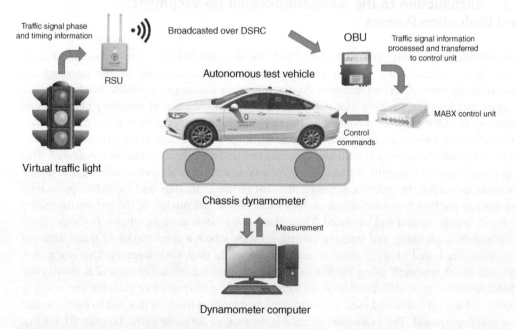

Figure 3.1 Illustration of DIL simulation. Source: Harman International Industries, Inc.; Denso Corporation; dSPACE GmbH.

to the driver steering wheel, that also turns with the actuator. This is a high-fidelity method of testing autonomous vehicle path-tracking control systems with the vehicle and its real steering actuator in the loop. To make this DIL/VIL testing more realistic, robotic vehicles are fit with shells corresponding to chosen vehicle types to generate relative motion profiles [6]. This approach is used for testing ADAS systems and cooperative driving in a large indoor test area. The highest level of in-the-loop simulation is when the vehicle is actually running in a large empty area, and the environment and traffic are emulated in real time such that all sensor data is generated and sent to the vehicle computers. This is not only Vehicle-in-Virtual-Environment (VVE) testing and is the most flexible and realistic but also safe testing of autonomous driving algorithms before an actual public road deployment. Re-creations of actual city blocks are also used to test autonomous vehicles in a controlled test environment [7]. While this approach is very useful, it suffers from the fact that the city block(s) environment used is fixed and cannot be easily changed. The testing in the actual controlled city block sometimes uses augmented reality for virtual traffic generation [8]. A safe and relatively cheaper way of testing cooperative driving algorithms is to use Vehicle-in-Virtual-Platoon or VVP testing in which the vehicle runs preferably in a closed circuit test area where the other vehicles in the platoon are emulated by their DSRC or cellular communicated information [9]. This approach can also be used to safely test collision avoidance algorithms and fuel economy algorithms based on vehicle automation and connectivity. A road-side unit (RSU) modem is also mounted on top of the test vehicle and emulates connected traffic lights for validating pass-at-green and eco-glide algorithms [2].

The final deployment steps that follow model or X-in-the-loop (also called XIL) where X stands for the different approaches to HIL simulation presented above are testing with a real vehicle or vehicles in a controlled test environment like a proving ground followed by public road testing. All these different simulation and road-based testing approaches are illustrated graphically in Figure 3.2 as a V diagram. Details of these approaches will be described further in the chapter.

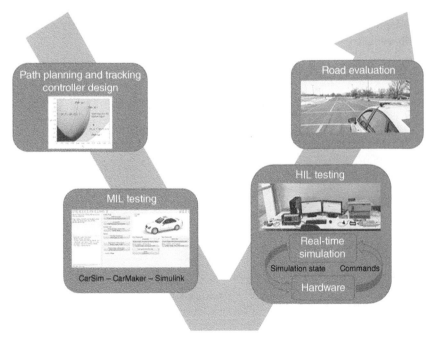

Figure 3.2 V diagram for model, XIL, and road testing of autonomous driving algorithms.
Source: Mechanical Simulation Corporation.

3.2 Model-in-the-Loop Simulation

This section is on MIL simulations and starts with the low- and high-fidelity vehicle models used in these simulations which are run offline. The virtual environments used in the MIL simulations are presented, starting with how the road network used for the autonomous vehicle path is generated. The Unity and Unreal game engine rendering environments and the corresponding publicly available LGSVL and CARLA simulators are then explained along with some of their interesting features useful for simulating autonomous vehicles.

3.2.1 Linear and Nonlinear Vehicle Simulation Models

Linear and nonlinear vehicle models for path following were presented in Chapter 2. These models are first implemented as Simulink models as illustrated in Figure 3.3 for the linear model and Figure 3.4 for the nonlinear model. In this book, the validated parameters of the research-level autonomous vehicle of the Automated Driving Lab at the Ohio State University are used [10] in the examples. This vehicle is a fully instrumented, drive-by-wire 2017 Ford Fusion Hybrid. The MIL vehicle models in Figures 3.3 and 3.4 are Simulink models that run very fast. This helps in speeding up the path-tracking controller development and its simulation evaluation.

While Simulink simulation of the models presented in Chapter 2 are useful and while the nonlinear path following model with nonlinear vehicle dynamics with tire force saturation has more fidelity than the linear models, a higher fidelity model that also incorporates the neglected subsystems such as suspension, steering, braking, powertrain, and more realistic and dynamic weight distribution are still desirable for more realistic results. It is possible to build such higher fidelity models by extending the mathematical modeling in Chapter 2 as in references [11, 12]. A more common approach that results in even higher fidelity models is to use a commercially available vehicle dynamics modeling programs such as CarSim, CarMaker, or other similar programs [13–15]. These programs have extensions that can be used to model heavy duty vehicles also. This higher fidelity vehicle simulation model use is presented in Section 3.2.2.

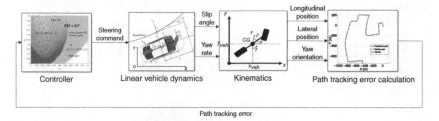

Figure 3.3 Linearized vehicle path following model.

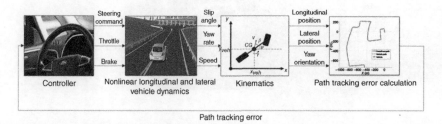

Figure 3.4 Nonlinear vehicle path following model.

3.2.2 Higher Fidelity Vehicle Simulation Models

CarSim is a widely used higher fidelity vehicle simulation model. It comprises of detailed models of all important subsystems of a vehicle which also means that a very large number of parameters and data in the form of tables need to be entered [13]. These values need to be validated against experimentally determined results of common vehicle dynamics tests such as coast down, full throttle speeding, step steering, ramp steering, constant radius turn, and swept sine steering and calibrated for the best fit. The companies providing these higher fidelity vehicle dynamics modeling tools like CarSim also offer a large number of predefined and built-in vehicle models in different vehicle classes. This means that the parameters are selected to represent the characteristics of the chosen vehicle class. Since these programs have a modular approach in which subsystem parameters can be entered separately, a commonly used approach by researchers is to start with a similar vehicle model template and make changes in the most important parameters like steering system parameters for path tracking, for example. Vehicle models in programs like CarSim use three-dimensional rigid body modeling of the main vehicle model and require the mass and moments of inertia to be entered. The vehicle is weighed on a scale and special test apparatus is used to determine the moment of inertia values for this purpose. If this is not possible, the best solution will be to use the parameters of the most similar predefined vehicle. The screenshot of a CarSim main screen showing the combination of subsystems for a generic E-class sedan is displayed in Figure 3.5 where summary information about the vehicle body such as mass and its aerodynamic drag, powertrain, steering, front and rear suspensions and tires is also shown. The suspension parameters are very important and they greatly affect handling behavior of the vehicle. A suspension test machine is used to determine these suspension system parameters experimentally.

CarSim also has sensors and traffic capabilities. The sensors are ground truth sensors emulating a localization sensor and perception sensors as shown in Figure 3.6, such as CarSim simulation environment with traffic, environment, and sensors. The localization sensor provides ground truth data on the absolute location of the vehicle. The perception ground truth sensor determines the exact location of nearby vehicles during simulations. It acts like the processed object data coming from an intelligent camera or radar sensor. It is possible to add noise to these sensor outputs to investigate the effect of sensor errors on autonomous vehicle operation like path-tracking control and obstacle avoidance. There are also fixed and vehicle-fixed cameras that can be placed at different locations to obtain simulated environment camera view information like a front-view

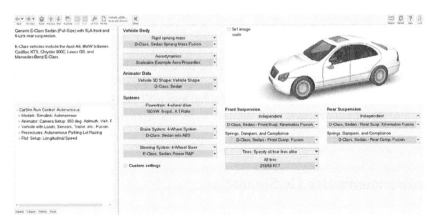

Figure 3.5 Generic E-Class sedan subsystem properties in CarSim. Source: Mechanical Simulation Corporation.

Figure 3.6 CarSim simulation environment with traffic, environment, and sensors.

camera [16]. Up 200 objects other than the simulated vehicle can be added to CarSim simulations. These objects usually represent other vehicles but can also be fixed objects like trees and buildings, vulnerable road users (VRU) such as pedestrians, animals, paint markings. If other vehicles are added to the simulation, they are kinematic objects whose motions are determined by the user. It is also possible to obtain the motion of these other vehicles from the cosimulation of a microscopic traffic simulator like SUMO for a reactive TIL simulation. Reference [17] has a similar cosimulation of CarMaker and the commercially available microscopic traffic simulator Vissim [18].

CarSim, CarMaker, and similar high-fidelity vehicle dynamics modeling programs can also be run in real time for HIL simulations. CarMaker has a physical sensors module [17] which can generate raw lidar data from the virtual environment used in the simulation. CarSim does not have that capability, but its vehicle dynamics can be immersed in the Unreal Engine rendering program and the user will have access to realistic autonomous vehicle sensors data from Unreal or a simulator like Carla that runs in Unreal while using realistic vehicle dynamics from CarSim. The current version of Matlab/Simulink also has a Vehicle Dynamics Blockset with a variety of models ranging from the bicycle model to higher-order models that can be used within the Unreal Engine environment including access to raw sensor data that can be processed using the algorithms in Matlab [19]. Matlab also has a lidar toolbox and a Robot Operating System (ROS) interface that can help with such realistic simulations [20]. For more detailed and realistic vehicle dynamics in Matlab, one needs to use the object-oriented programming in its Simdriveline and other similar modules [21].

3.3 Virtual Environments Used in Simulation

For MIL and HIL simulation and validation of autonomous driving functions and safety evaluation of a planned deployment using simulation tools, a realistic virtual environment is extremely important for several reasons that include the use of a correct map of roads, positions of objects in the environment, road infrastructure, and features in the environment and the ability to add traffic and VRU such as pedestrians and bicyclists. Different open-source platforms for building the simulation environment and running the autonomous driving simulations were studied and evaluated for simulation testing of low-speed autonomous shuttles in predetermined urban routes in geofenced areas in references [22, 23] for example. The process of creation and construction of the simulation environment for realistic autonomous-driving simulation is treated in this subsection. A recent autonomous shuttle deployment area within the Smart City Challenge winner Smart Columbus project of Columbus, Ohio, US, is used in presenting this approach. The geofenced urban area where these autonomous shuttles have operated is the Linden Residential Area.

To ensure the accuracy of the map and repeatability of results and scalability to other locations of autonomous driving simulation evaluation, the open source and freely available map called OpenStreetMap (OSM) was used for creating the simulation environment road geometry. For creating objects in the environment of the map and road infrastructure, multiple rendering algorithms were used for creating the corresponding meshes and object surfaces.

3.3.1 Road Network Creation

Currently, many maps are publicly available for everyday use such as Google Maps or Bing maps. High-definition maps that are specifically built for autonomous driving are also available for highways from companies like HERE and TomTom. In contrast to these commercial maps, the OSM is a freely available, open-source map. Thus, the users have access to the geodata underlying the map. The data from OSM can be used in various ways, including production of paper maps and electronic maps (similar to Google Maps, for example), geocoding of address and place names, and route planning. As it is open source, it is editable by using different scripts. The road network in the OSM map is relatively complete so that it can also be used for navigation purposes and in traffic simulations. Since OSM is based on user contributed data, there are sometimes minor problems that need to be fixed before using it to create the road network in a simulator. Possible problems include an incorrect number of lanes, incorrect intersection geometry, and outdated speed limits. Before using OSM, the map has to be updated and shared with the rest of the user community. In rural, residential areas, significant objects such as residential housing, buildings, and some road signs are not modeled and mapped. This, of course, is due to the map not being up to date and causes a lack of environmental information which affects the realism of creating the autonomous driving simulation environment based on the OSM map. Thus, using scripts such as the Java OpenStreetMap (JOSM) editor, the OSM map information is improved by updating incorrect or missing information and adding complete housing information along the chosen area of interest. This was done in [23], for example for the Linden Residential Area so that it can be used in simulation environment development as shown in Figure 3.7. While this procedure requires a lot of manual editing, once the updated OSM is pushed to the OSM server to update the source map, it becomes available to the whole user community of OSM maps.

The program called RoadRunner is a very useful tool for editing the road network created using the above or a similar approach [24]. RoadRunner makes it very easy to interactively add roads that are drivable in simulation, with lane markings, roundabouts, infrastructure, lanes, and the like in order to create three-dimensional scenes based on the actual road geometry for simulation of autonomous driving. RoadRunner used to be a separate, commercially available program but is part of Matlab/Simulink since 2020. It can still run as a standalone product. The available library of road and infrastructure assets makes it very easy to go from a road network to a version that

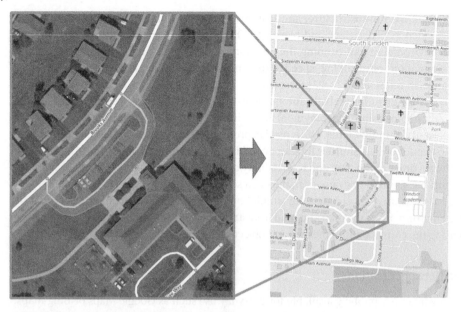

Figure 3.7 OpenStreetMap editing and exporting.

can be used in simulations in a large number of existing simulators. It is also possible to use GIS tools to import aerial imagery and point cloud data into RoadRunner created environments. Simulation environments created or enhanced in RoadRunner can be saved in a large number of formats including the filmbox format for *.fbx files. These environments can be used in both the Unity and the Unreal game engine simulations and freely available autonomous driving simulation programs like LGSVL and CARLA and a lot of similar ones. Indeed, RoadRunner is one of the few road editing programs that can create simulation roads for use with the popular autonomous driving simulation program CARLA. Roads and intersections created using RoadRunner for the Linden Residential Environment for use in CARLA simulations is shown in Figure 3.8 as an example.

Figure 3.8 Some roads and intersections in Linden Residential Environment created using RoadRunner.

3.3.2 Driving Environment Construction

Unity Engine

The Unity platform is a cross-platform game engine developed by Unity Technologies for real-time, three-dimensional graphics visualization. This engine has been adopted in many areas outside the field of video games such as filming, automotive CAD, architecture, and construction. With the rise of the autonomous vehicle industry, the Unity platform is now applied more and more in the field of autonomous driving. The Unity platform is one of the platforms that can be used for visualization of the autonomous driving environment and for extracting autonomous vehicle raw perception sensor data. The freely available, open-source LGSVL simulator [25] is based on the Unity platform.

One starts with the OSM map discussed in Section 3.3.1. Various Unity scripts and assets are used for importing the OSM environment. Freely available and low-cost Unity asset libraries can be used for this purpose. The imported map is going to have complete environment information containing the road network, traffic infrastructure, buildings, and other related objects. The pipelined approach shown in Figure 3.9 is used for this purpose.

After editing the OSM map according to available satellite images and height images, the map is then imported into the Unity Engine. At this stage, the map in Unity contains the road network and the raw mesh without any texture which looks different from the corresponding real-world appearance as shown in Figure 3.10. In a subsequent step, road signs and road features are also added and rendered in the Unity Engine so that the real-world environment can be completely emulated in the simulations. Those objects are usually rendered manually by comparing with other objects in the simulation environment and camera images or video screenshots.

Different methods for rendering the meshes can be used to better represent the real-world objects as shown in Figure 3.11. Different mesh-generating methods can be used for mesh generation and rendering like the MeshLab [26] asset in the Unity Asset Store. The Easy Road 3D Asset Package [27] in the Unity platform can be used to render the road surface. Default meshes available in the Unity project for surface rendering can be for the buildings. An autonomous vehicle simulation running in the Unity environment created using the procedure outlined in this subsection within the LGSVL simulator is shown in Figure 3.12 where the freely available Autoware [28] autonomous driving software is used for path tracking.

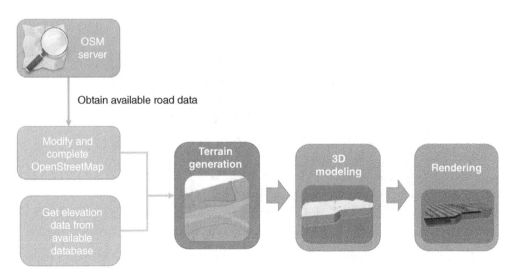

Figure 3.9 Pipelined approach for importing an OSM map into the Unity Engine.

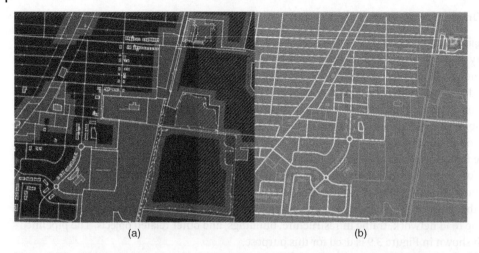

Figure 3.10 OpenStreetMap (a) and corresponding unedited, imported map in Unity (Linden Residential Area) (b).

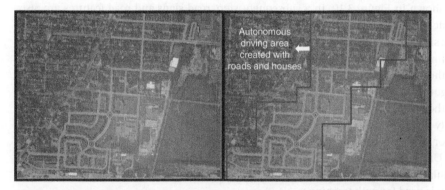

Figure 3.11 Three-dimensional model of the simulation environment where road surface and road features are modeled realistically.

Figure 3.12 LGSVL simulator autonomous vehicle (a) controlled by Autoware (b) operating in Unity engine simulation environment.

Unreal Engine

Similar to Unity, Unreal is another game engine-type graphics-rendering platform that is used in autonomous driving simulation environment building. The freely available and open-source CARLA simulator [29] uses the Unreal Engine for its graphic visualization of the simulated environment. The pipelined approach that can be used for creating the Unreal Engine simulation environment shown in Figure 3.13 is discussed in this subsection in the context of the Linden Residential Area route. The overall aim is to use time-stamped three-dimensional (3D) lidar data to reconstruct the surface of the entire environment and then to add texture details from Google Maps (Google Earth Studio [30]) and road details from OSMs.

The point cloud data is reduced first to increase the processing speed without considerable loss in environment reconstruction. Point cloud data density depends on measurement distance due to the lidar construction with sensor channels projecting laser signals being at fixed angular separation. As a result, the point cloud will be very dense near the AV and sparse at further distance. It is possible to reduce the lidar point cloud data about 40% by using the Poisson Sampling Disk approach to remove highly cluttered point clouds using sampling disks as these usually correspond to other vehicles on the road. Such unwanted point clouds that correspond to objects that are stationary for a significant time during recording, for example any stationary car waiting for a traffic signal should be removed using the SqueezeSeg [31] Convolutional Neural Network (CNN) or other similar methods. The freely available and open-source MeshLab, Blender [32], and Point Cloud Library (PCL) [33] tools can also be used during preprocessing. Intermediary steps like feature extraction, point-to-point correspondence, and feature matching can be easily taken care of using tools from the freely available, open-source Meshroom [34] program.

For the Linden Residential Area example illustrated in Figure 3.14, nearly 4000 satellite images of the Linden Residential Area route were retrieved from Google Earth Studio [30] and used for generating mesh texture [22, 23]. It is also possible to collect data using drones, i.e. aerial imagery,

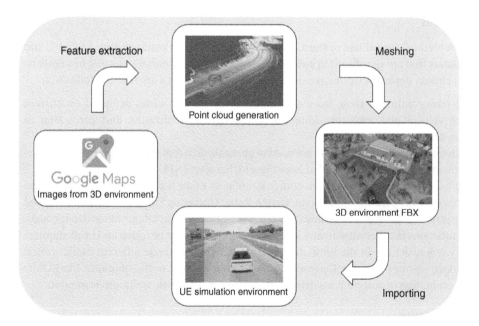

Figure 3.13 Pipelined approach for importing map data into the Unreal Engine environment. Source: Google LLC.

Figure 3.14 Local features of the mesh of the Linden Residential Area.

in order to construct a more realistic and up-to-date replication of the environment. It is not possible to recover texture information in occluded regions such as those under the trees. The usual procedure is to drive an instrumented vehicle with several cameras and high-accuracy GPS and use the collected images to create the environment rendering.

3.3.3 Capabilities

Some of the important capabilities of the Unity and Unreal Engine platform-based LGSVL and CARLA simulators that are significant in autonomous vehicle (AV) simulation testing in a realistic rendering of a chosen deployment site like the Linden Residential Area are listed as follows:

- The virtual sensor suites in these simulators emulate the sensor suites deployed on current autonomous vehicle platforms, providing the capability of localization and perception in autonomous-driving simulation.
- Multiagent features in these simulators are used to generate different traffic scenarios of interaction with other vehicles which are called Non-Player Character (NPC).
- The easy-to-use interfaces for these open-source simulators make it possible to connect them to external software and operating systems such as the Robot Operating System (ROS) or ROS-based autonomous driving systems like Autoware so that readily available autonomous driving algorithms and functions in different software and operating systems can be tested and implemented.
- Additionally, it is easier to edit the simulation environment and generate different traffic scenarios in these open-source platforms. Cosimulation with a microscopic traffic simulator-like SUMO enables the evaluation of autonomous driving algorithms in a realistic traffic environment.

Unity – LGSVL Simulation
LGSVL is a Unity-based freely available and open-source simulator [25]. Its latest version has out-of-the-box integration with the open-source Apollo platform and the Autoware platform for

Figure 3.15 Unity-based LGSVL simulator emulation of 3D lidar and camera sensors.

autonomous driving and can emulate 128 channel 3D lidar along with camera and radar sensors. The other virtual sensors used in the LGSVL simulator are GPS and Inertial Measurement Unit (IMU) of the AV in the simulation. The 3D lidar point cloud emulation of the LGSVL simulation in the Unity platform and several types of real-time camera images is shown in Figure 3.15. During the simulation with a computer with Central Processing Unit (CPU) and Graphical Processing Unit (GPU), most of the ray casting-based lidar computations are handled by the GPU while the distance information is calculated in the CPU. The lidar points can be seen projected on the environment as dots. A wider view of the emulation of 3D lidar point cloud data and the projection of the points on the environment is shown in Figure 3.16.

The Unity-based LGSVL simulation cameras render pixels within the preset range and provides Red-Green-Blue (RGB) color, depth, or semantic information per pixel in the camera data which is illustrated in Figure 3.17. Besides the functionality of emulating essential AV sensors, the Unity Engine-based simulation environment has the capability of simulating NPCs which are used to model the other vehicles on the road. These NPC vehicles follow predefined lanes and obey traffic rules. It is possible to investigate the operation of the AV within traffic in the LGSVL simulator using the NPC vehicles. The traffic added to the environment was shown along with the radar sensor detections in Figure 3.18. There are short-range wide-angle and long-range narrow-angle radar sensors on the simulated autonomous vehicle. These radar sensors detect vehicles (can be seen as white lines in the figure) and provide relative position information.

Figure 3.16 Three-dimensional lidar point cloud emulation reflected on environment in Unity-based LGSVL simulation.

Figure 3.17 Depth, RGB, and semantic segmentation cameras in Unity-based LGSVL simulator.

Figure 3.18 Radar sensor detections with simulation of traffic using NPC vehicles in Unity.

Figure 3.19 CARLA AV simulation in the Linden Residential Area.

Unreal Engine – CARLA

The freely available and open-source CARLA simulator is used with the Unreal Engine platform simulation environment. CARLA has a powerful Application Programming Interface (API) that allows users to control all aspects related to the simulation including traffic generation, pedestrian behavior, weather conditions, and sensors. CARLA also has its built-in autonomous driving function which is very useful in simulations used for collecting training data. Users can configure diverse sensor suites including lidars, multiple cameras, depth sensors, and GPS for the AV. Within the simulation, the ego vehicle's (AV) state consisting of its speed, heading angle, and global location are easily seen using the virtual GPS sensor along with the states of the vehicle actuators. These synthetic data can be accessed from other programs or software through the APIs provided by the CARLA simulator. Figure 3.19 shows the CARLA simulation of an AV operating in the Linden Residential Area environment. Figure 3.20 shows the corresponding camera depth map and Figure 3.21 shows a 3D lidar scan in CARLA.

Figure 3.20 Depth camera display in the CARLA simulator.

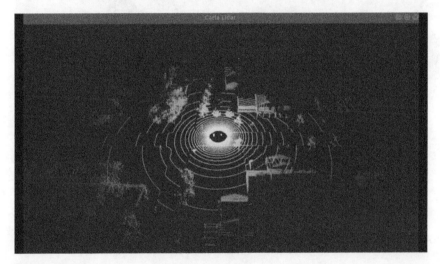

Figure 3.21 3D lidar scan in the CARLA simulator.

3.4 Hardware-in-the-Loop Simulation

The HIL simulator used in the examples in this book is a state-of-the-art simulator that is aimed to realistically test controllers and algorithms before implementation on a real vehicle [2]. Software used consists of mainly CarSim and Simulink. The main hardware that is in the loop in the simulator is a dSPACE MicroAutoBox (MABx) electronic control unit which runs the autonomous vehicle path-tracking control and collision-free maneuvering computations. The vehicle is emulated by the validated CarSim model running in real time in the dSPACE SCALEXIO real-time vehicle model simulation computer. Other hardware that is used in some examples includes an in-vehicle PC with high-performance GPU for running the perception and decision-making algorithms in real time, two Denso on-board units (OBU) for vehicle connectivity, and a Savari RSU connected to an Econolite traffic control unit for infrastructure connectivity. The MABx electronic control unit behaves as if it is connected to a real vehicle, receives sensor information or information

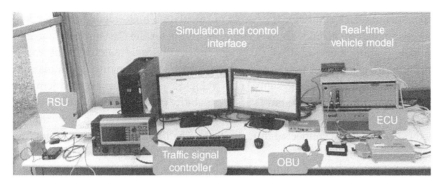

Figure 3.22 HIL simulator components.

about the vehicle, while running the designed controllers and algorithms to calculate and send steering, throttle, and brake signals to actuate the virtual vehicle, all through the CAN bus as in the real vehicle. The SCALEXIO vehicle model simulation computer receives these control signals and uses them as inputs for the vehicle dynamics calculations of the virtual vehicle running in the real-time simulation. The CarSim model is cosimulated as the autonomous vehicle within the Unreal Engine or within CARLA with raw sensor data feedback for more realistic HIL simulation of autonomous driving. It is also possible to cosimulate a microscopic traffic program like SUMO for realistic TIL simulation. The same computer connected to the SCALEXIO or several computers may be used for the Unreal-CARLA and SUMO simulations. The HIL simulator elements and communication between HIL elements are illustrated in Figure 3.22. A similar architecture can be built using real-time CarMaker with its physical sensor models for raw sensor data generation and VisSim interface add-ons to simulate realistic TIL simulations completely within CarMaker [17].

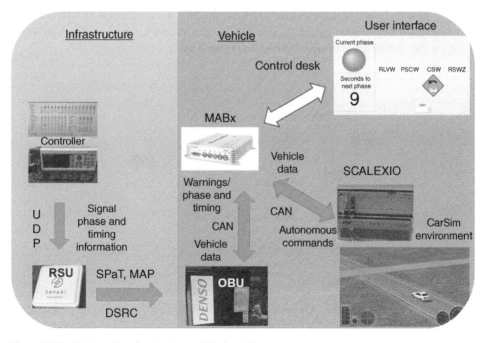

Figure 3.23 Information flow between HIL simulator components.

With these components, the HIL simulator is capable of calculating high-fidelity autonomous vehicle dynamics in real-time. Moreover, with the parameter tuning for the CarSim vehicle using the experimental vehicle's parameters, the HIL simulation results very closely match the real-world experimental results. CarSim also allows using various kinds of scenario elements for simulating different scenarios, such as traffic lights, road signs, other vehicles, and pedestrians as kinematic objects. The Vehicle-to-Everything (V2X) communication capability of the HIL simulator through its DSRC radio devices (OBU and RSU) allows HIL simulations of connected vehicle applications including connected pedestrians and connected infrastructure. The information flow diagram for HIL simulation is shown in Figure 3.23.

With the current capabilities and realistic simulation environment, the HIL simulator provides a significant advantage for testing numerous scenarios repeatedly and doing adjustments and optimizations on controllers or algorithms before implementing them in a real vehicle and conducting real experiments. After those adjustments, the resulting controllers or algorithms can be used on the real vehicle with very few or no modifications. Testing in the virtual environment is also much quicker and safer.

3.5 Experimental Vehicle Testbeds

For most of the experimental results presented in this book, a drive-by-wire Ford Fusion hybrid electric vehicle with automated driving capability was used. A smaller drive-by-wire Dash EV electric vehicle was also available in testing that required either a smaller vehicle or multiple vehicles. These vehicles were automated using the unified architecture of reference [35] that includes multiple types of sensors for perception and localization along with computers for processing and units. The idea behind this unified architecture is to reduce the effort that goes into designing a new architecture for every vehicle that needs to be automated and sometimes also to easily move the automation system from one vehicle platform to another. The hardware configuration for the unified architecture is shown in Figure 3.24.

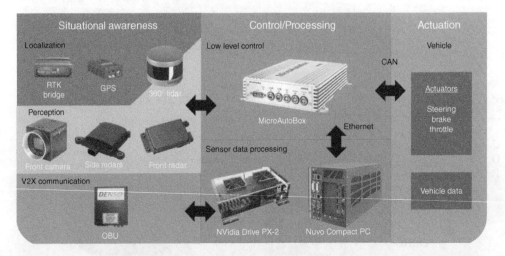

Figure 3.24 Unified AV architecture. Source: Velodyne Lidar, Inc.; Teledyne FLIR LLC; BorgWarner Inc.; Denso Corporation; dSPACE GmbH; NVIDIA Corporation; Neousys Technology.

3.5.1 Unified Approach

In this section, we present the hardware library and software library of the unified architecture. An autonomous vehicle needs sensors to perceive its environment, processing units to process the data from these sensors and actuators to drive the vehicle without a human driver. The main idea of the unified architecture is to design an autonomous vehicle structure that can be easily modified or adapted and implemented on another vehicle platform regardless of size and type to achieve replicable and scalable autonomous driving. After the implementation, only necessary modifications to optimize the performance would be tuning the low-level controller parameters according to the specific vehicle platform used and fine tuning of the autonomous driving functions for the new platform [23, 35].

As seen in Figure 3.24, several different types of sensors are included in the design to overcome shortcomings of each other. An OBU is also added to provide DSRC communication capabilities so that the vehicle can overcome non-line-of-sight (NLOS) situations and V2X applications can be implemented to provide cooperative collision prevention functionality. It is also possible to add more communication technologies such as Bluetooth and Cellular to increase the capability for V2P communication. The main localization sensor is a real-time kinematics (RTK) GPS sensor with built-in IMU and Kalman filter GPS/IMU sensor fusion. This sensor has very high accuracy for global position measurements. The accuracy is up to around 50 cm for normal use and can go up to 2 cm when RTK corrections are used, which are obtained through cellular connection by the RTK bridge. Along with the GPS sensor, a Lidar sensor(s) is also used for localization with algorithms such as simultaneous localization and mapping (SLAM) [36] and 3D mapping/map matching [22, 23] to localize the vehicle in the event of GPS signal loss. The autonomous vehicle automated with this architecture and used in the experiments in this book is shown in Figure 3.25 with the location of its AV sensors and computers.

Path-following control algorithms are studied in detail in subsequent Chapters 5–7. Here, some of the basics of the parameter space approach for autonomous driving that is part of the unified framework presented here will be discussed [37, 38]. The first step is to develop the Simulink implementations of the longitudinal and lateral vehicle dynamics models that are parametric with respect to the chosen significant parameters as presented in Chapter 2. Then, the parameter space methods of D-stability, phase margin, and weighted sensitivity bound constraints as illustrated in Figure 3.26 can be used in the longitudinal and lateral controller designs for trajectory tracking. An uncertainty box is used to characterize the different road friction coefficient and vehicle speed ranges which are different for the two different-sized vehicles treated in this book as shown in Figure 3.27.

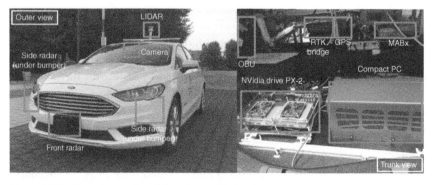

Figure 3.25 Experimental vehicle using unified architecture.

Multi-objective design

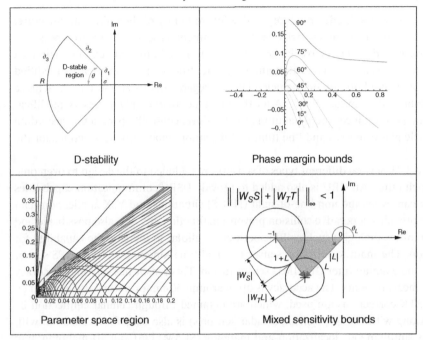

Figure 3.26 Unified multiobjective parameter space approach for trajectory following controller designs.

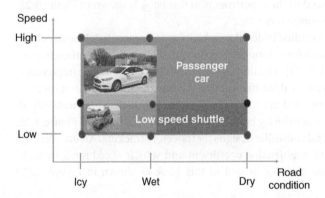

Figure 3.27 Uncertainty boxes of operating conditions for robustness analysis.

Vehicle mass and speed ranges are shown in a similar fashion for these two vehicle types in Figure 3.28. Parameter space control design methods are used to achieve similar performance of the low-level controllers for these two different categories of vehicles by scaling the design based on calibratable parameters.

3.5.2 Unified AV Functions and Sensors Library

The unified AV functions and sensors software library structure designed in Simulink is shown in Figure 3.29. This Simulink software library is designed under several main titles. Localization blocks are used to receive localization data from different sensors where some of them are

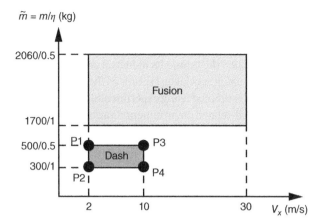

Figure 3.28 Vehicle mass and speed uncertainty regions for two categories of vehicles. Source: Zhu et al. [39].

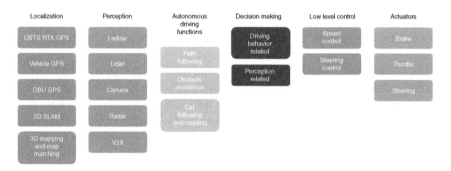

Figure 3.29 Unified library of AV functions and sensors.

processed and coming from processing units while others are directly measured and transferred to the electronic control unit. Similarly, perception blocks are used to receive perception data from sensors or processed data from their processing units. Autonomous behavior algorithm blocks make necessary computations for autonomous driving algorithms such as path following and obstacle avoidance for lateral driving and car following and cruising for longitudinal driving. Decision-making part has blocks that use all the information available to the computers such as localization and perception to determine the autonomous driving state in a finite-state machine implementation of decision-making. Deep-learning algorithms used for decision-making are also under that heading. Low-level control blocks are for calculating throttle, brake, and steering commands required for executing the autonomous driving functions determined by decision-making. Lastly, actuator blocks are used to transfer outputs from these low-level controllers to the real actuators inside the vehicle through communication such as CAN or analog signals.

This design was implemented with the hardware architecture and library part in the two different-sized vehicles used in this book for autonomous path following and collision-free maneuvering experiments. While the sedan vehicle shown in Figure 3.25 is aimed more for moderate to high-speed autonomous driving applications, the smaller neighborhood electric vehicle shown in Figure 3.30 is aimed for lower-speed autonomous shuttle-type driving applications.

3.6 Estimation

Some parameters of vehicle models like lengths and mass are published in technical specifications by the manufacturer. There are other parameters like the moments of inertia and suspension parameters that need to be determined experimentally using special test setups as these are usually not shared. There are also some parameters that can be estimated using experimentally determined responses of the vehicle to test inputs. There are also road-related environmental variables that need to be estimated using a similar approach. This procedure is presented in this section using effective tire radius and tire road friction coefficient estimation as examples.

3.6.1 Estimation of the Effective Tire Radius

The effective tire radius refers to the distance from the axle center of the tire to its contact patch with the ground. Due to the load applied, its value is generally smaller than the free-load tire radius and depends on the tire type and other tire dimensions. The effective tire radius also depends on other factors such as tire tread wear condition, the inflation pressure, and change of vehicle load. Though the effective tire radius changes encountered in practice might be slight, they could affect the estimation of the slip ratio, the accuracy of which is important for road friction coefficient estimation. It is, hence, desirable to have the effective tire radius estimated before the calculation of slip ratios.

Several methods are available to choose from the literature as in [40] and the main idea is similar in which the measured/estimated information of the tire rotational motion and its linear motion along its longitudinal direction are related to each other. The tire rotational speed can be measured with the encoder sensor mounted at each wheel that is already available for any vehicle equipped with the mandatory anti-braking system (ABS). The linear motion in the tire longitudinal direction is more difficult to be measured or estimated. Typical methods that can be used to derive it include the longitudinal acceleration measured from the accelerometer or the linear velocity estimated with methods such as the extended Kalman filter.

The tire longitudinal speeds v_{fl}, v_{fr}, v_{rl}, and v_{rr} corresponding to the different wheels can be obtained from the vehicle longitudinal speed v_x and lateral speed v_y as

$$v_{fl} = \left(v_x - \frac{W}{2}r\right)\cos\delta_f + (v_y + l_f r)\sin\delta_f, \tag{3.1}$$

$$v_{fr} = \left(v_x + \frac{W}{2}r\right)\cos\delta_f + (v_y + l_f r)\sin\delta_f, \tag{3.2}$$

$$v_{rl} = v_x - \frac{W}{2}r, \tag{3.3}$$

Figure 3.30 Dash EV experimental vehicle using unified architecture. Source: Velodyne Lidar, Inc.; Teledyne FLIR LLC; LeddarTech Inc.

$$v_{rr} = v_x + \frac{W}{2}r, \tag{3.4}$$

where W is the track width, l_f and l_r are distances from the vehicle center of mass to the front and rear axles in the bicycle mode, respectively, and r is the vehicle yaw rate. The relation

$$v_{ij} = R_{eff,ij}\omega_{ij}, \quad i = f, r \; j = l, r, \tag{3.5}$$

where $R_{eff,ij}$ refers to the effective tire radius for the ij th wheel is only true if the tire does not slip on the road surface. This ideal condition will not be met in practice as slip is needed for tire forces to be generated. Large slip occurs when the wheel is being driven with sudden traction torque or being slowed down with sudden braking torque. The tire slip is insignificant and can generally be ignored when the wheel is passive, i.e. in the absence of either drive or brake torque application.

The recursive least square (RLS) algorithm is suitable for the linear regression problem (3.5). A variation of the standard RLS algorithm is used here that emphasizes recent data more in the estimation results by introducing an exponential forgetting factor λ $(0 < \lambda < 1)$ in the sum of square errors

$$V\left(\hat{\theta}, t\right) = \frac{1}{2}\sum_{i=1}^{t}\lambda^{t-i}\varepsilon^2(i) = \frac{1}{2}\sum_{i=1}^{t}\lambda^{t-i}\left(y(i) - \varphi^T(i)\hat{\theta}\right)^2 \tag{3.6}$$

to be minimized.

In this way, old data are given less and less weight as the iterations continue over time until finally they become negligible. The changing of the effective tire radius due to tire condition change (tire wear, inflated pressure, load, etc.) can be sensed in time with the introduced forgetting factor. The RLS iteration with the forgetting factor λ is expressed as [41]

$$\hat{\theta}(t) = \hat{\theta}(t-1) + K(t)\left(y(t) - \varphi^T(t)\hat{\theta}(t-1)\right), \tag{3.7}$$

$$K(t) = P(t-1)\varphi(t)(\lambda + \varphi^T(t)P(t-1)\varphi(t))^{-1}, \tag{3.8}$$

$$P(t) = (I - K(t)\varphi^T(t))P(t-1)/\lambda. \tag{3.9}$$

3.6.2 Slip Slope Method for Road Friction Coefficient Estimation

The forces between the tires and the road are very significant for vehicle dynamic analysis as they provide the means of controlling the motion of the vehicle. The vehicle operations, including speeding up, braking, and steering, all use the changing of the tire-road forces to realize these maneuvers. The maximum tire force is hence of great interest as it determines the vehicle handling limits and stability.

To show how the value of the road friction coefficient is determined, the normalized force for each tire is introduced first as follows:

$$\rho_{x,tire} = \frac{F_{x,tire}}{F_{z,tire}}, \tag{3.10}$$

$$\rho_{y,tire} = \frac{F_{y,tire}}{F_{z,tire}}, \tag{3.11}$$

where $F_{x,tire}$, $F_{y,tire}$, and $F_{z,tire}$ are the tire longitudinal, lateral, and vertical/normal force, respectively. The normalized tire traction force, also named as the coefficient of traction, is the resultant of the two in (3.10) and (3.11) given by

$$\rho = \sqrt{\rho_{x,tire}^2 + \rho_{y,tire}^2}. \tag{3.12}$$

The road friction coefficient μ is defined as the maximum normalized traction force the tire could supply. Its value varies between 0 and 1 depending on the road surface type (such as asphalt and gravel) and condition (dry, wet, or icy).

Knowledge of this coefficient is, hence, necessary for safe maneuvers for vehicle motion planning and control. Direct sensor measurements are available with options ranging from acoustic sensors, strain sensors vulcanized in the tire tread, to optical sensors relying on ground reflection [42–45]. However, equipping the vehicle with these extra sensors adds the unnecessary problems of sensor reliability and robustness under different driving conditions. Cost is also a factor that prevents their application in series produced vehicles. These are the main motivations for the use of estimation techniques.

The slip-slope method which takes advantage of the tire longitudinal characteristics is used here for estimation. The slip ratio of the tire is defined to measure the relative difference between the tire's circumferential velocity and its linear velocity referenced to the ground as follows:

$$s_x = \frac{R_{eff}\omega - v_{tire}}{\max\{R_{eff}\omega, v_{tire}\}}, \tag{3.13}$$

where ω is the wheel rotational speed and v_{tire} is the tire longitudinal speed that can be approximated from Eqs. (3.1)–(3.4).

Figure 3.31 shows the typical relationship between the normalized longitudinal force and the tire slip ratio using the tire magic formula for different road friction coefficient values. It is observed that the longitudinal force first increases with the slip ratio until it reaches a peak whose value depends on the road type and then ramps down a little bit in a saturation like behavior.

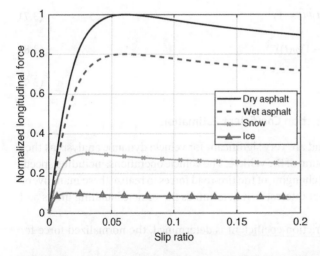

Figure 3.31 Normalized longitudinal force as a function of slip computed using magic tire formula model.

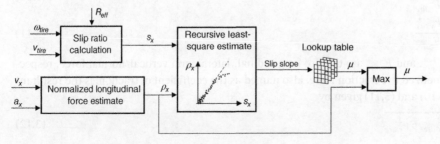

Figure 3.32 Estimation diagram of the slipslope method.

If the slip ratio is large such that the peak of the normalized longitudinal force is already observed, then the problem solution is straightforward and is obtained by simply choosing the peak value as the road friction coefficient according to the definition in Eq. (3.12). However, the slip ratio is generally small in daily driving, meaning such method is only applicable in rare occasions such as slipping on icy road. Fortunately, the longitudinal force appears to have a nice linear relationship with the longitudinal slip at small slips, and the slope differs for different road surfaces. If the slope value could be identified, then the corresponding road friction coefficient could also be determined. This idea forms the backbone of the so-called "slip-slope" method. The process of the slip-slope method is illustrated as shown in Figure 3.32 where a lookup table induced from Figure 3.31 is needed to relate the slip slope to the road friction coefficient.

At small steering angles, the total longitudinal force from all tires can be estimated from the vehicle longitudinal dynamics given by

$$F_x = m\dot{v}_x + mg \sin \theta + F_{roll} + F_{air}, \tag{3.14}$$

where m is the total vehicle mass, $\dot{v}_x$ is the longitudinal acceleration, $mg \sin \theta$ is the term accounting for the road grade effect with θ being grade angle, $F_{roll} = C_{roll} mg \cos \theta \approx C_{roll} mg$ is the rolling resistance force with C_{roll} being the rolling resistance coefficient, and $F_{air} = (1/2)\xi C_d A v_x^2$ with ξ being the ambient air density, C_d the aerodynamic drag coefficient, A the frontal projected area, and v_x the longitudinal vehicle speed. Note that the accelerometer reading in the longitudinal direction $a_{x,measured} = \dot{v}_x + g \sin \theta$ already includes the road-grade effect, making the first two right-hand-side terms in (3.14) readily available.

For the slip ratio calculation, both the tire rotational speed and the effective tire radius could achieve high accuracy with either on-board sensors or estimation, while accuracy of the tire longitudinal speed depends heavily on the vehicle longitudinal speed with Eqs. (3.1)–(3.4). An estimation error of merely 0.05 m/s in the vehicle longitudinal speed could lead to error of the slip slope by up to 0.005 at around 10 m/s. This accuracy is enough for ABS or ESC operation since they mainly operate at large slips but is not favorable for the slip slope method since the slip slope of interest is within very small slip regions. In the case that the vehicle longitudinal speed accuracy could not be ensured, an alternative way is to take advantage of the rotational speed measurements of the passive wheels. The slip of passive wheels is small enough to be omitted during coasting and acceleration phase. The tire longitudinal speeds for the front and rear wheels at the same side (left/right) can be assumed equivalent at small steering, as observed from Eqs. (3.1)–(3.4). It is, hence, reasonable to approximate the tire longitudinal speeds at the front two driven wheels with

$$v_{fl}[k] \approx v_{rl}[k] \approx R_{eff}[k]\omega_{rl}[k], \tag{3.15}$$

Figure 3.33 Experiment road condition.

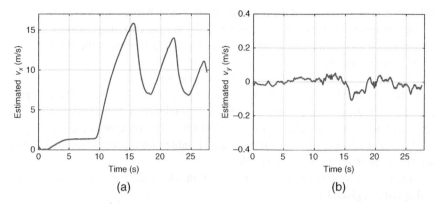

Figure 3.34 Vehicle longitudinal (a) and lateral (b) speeds estimated with the EKF.

$$v_{fr}[k] \approx v_{rr}[k] \approx R_{eff}[k]\omega_{rr}[k], \tag{3.16}$$

where $R_{eff}[k]$ is the online estimated tire radius value at time k using RLS as stated in Section 3.6.1, $\omega_{rl}[k]$ and $\omega_{rr}[k]$ are the rear wheel speed measurements. Notice that the above equations are only valid during acceleration for the front-wheel-driven vehicle. Similar assumptions and approximate equations can be developed for rear-wheel-driven or all-wheel-driven configurations as long as not all wheels are being driven at the same time. Otherwise, Eqs. (3.1)–(3.4) could be used for slip-ratio calculation.

3.6.3 Results and Discussion

Experiments on Dry-Asphalt Road
The road experiment data collected with our Ford Fusion test vehicle on dry asphalt is used for the remaining analysis of the effective tire radius estimation and the road friction coefficient estimation. The road surface is not quite flat as shown in Figure 3.33. The vehicle is equipped with a dual-antenna GPS (OXTS xNAV550) with RTK correction for localization and a steer-by-wire system. The vehicle repeats acceleration movements during the test. Its longitudinal and lateral velocity estimates from an extended Kalman filter by fusion of the GPS and IMU sensors are shown in Figure 3.34 at a sampling frequency of 100 Hz.

Since our experimental vehicle is driven by front wheels, it is reasonable to assume that the rear two tires do not slip during coasting or acceleration, based on which the relation (3.5) can be utilized for tire radius estimation. Measurements of all four tire rotational speeds are available from wheel sensors as shown in Figure 3.35. The tire longitudinal speeds for the rear two wheels are estimated with Eqs. (3.3) and (3.4), the data of which during acceleration are then drawn against the measured tire rotational speeds, as shown in the scatter plot in Figure 3.36. The linear relation is evident in the plot for the two rear tires.

The RLS updates the estimation of the effective tire radius during the acceleration phase at the sampling frequency of 100 Hz, with a high forgetting factor of $\lambda = 0.999$, and initially set as $\hat{\theta}(t_0) = 0.3$, $P(t_0) = 1 \times 10^4$. The estimation results for the rear-left and rear-right tires are shown in Figure 3.37. Both curves respond very quickly when the estimation starts and converge to the value around 0.321 m, which matches the mounted tire type P255/50R17 (free-load tire radius of about 0.34 m) on the test vehicle.

The normalized longitudinal forces ρ_x obtained with Eqs. (3.10) and (3.14) during acceleration are drawn against the average slip ratio for the front two tires $s_{x,f}$ estimated using

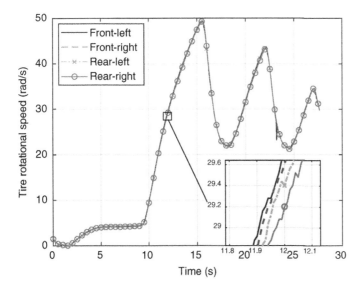

Figure 3.35 Measured wheel rotational speeds during the experiment.

Figure 3.36 Scatter plot of the rear tire longitudinal speed and its rotational speed during acceleration.

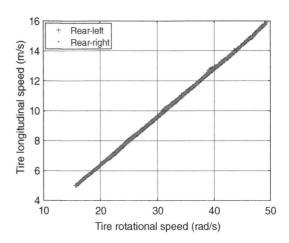

(3.13) and (3.15)–(3.17) in Figure 3.38. It is seen that the linear model

$$\rho_x = K_x s_{x,f}, \tag{3.17}$$

is a proper fit for the scattered data, where K_x is the slipslope with its least square estimate achieving the coefficient of determination around 0.8.

The RLS algorithm (3.7) and (3.8) is again suitable for such linear regression fit with the slip slope estimate shown in Figure 3.39. The RLS updates during the acceleration phase at the sampling frequency of 100 Hz with a high-forgetting factor $\lambda = 0.999$, and initials set as $\hat{\theta}(t_0) = 0$, $P(t_0) = 1 \times 10^5$ shows that the slip-slope converges quickly once RLS starts updating around 10 seconds and stays consistent afterward at around 43.9. The peak normalized force value during acceleration is around 0.83 from Figure 3.38 which means that the friction coefficient of the test road is at least larger than this value. This is reasonable as the experiment was conducted on a dry asphalt road surface with its road friction coefficient considered to be around 1.0. Although we have not shown the experimental analysis on other types of road surface, it is expected that the same methodology would work

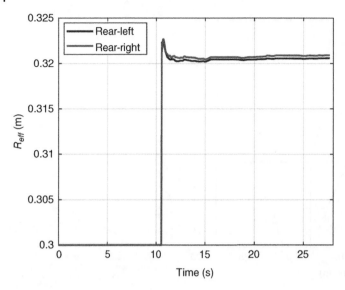

Figure 3.37 Effective tire radius estimated for the rear two wheels.

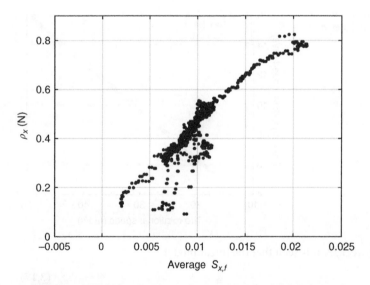

Figure 3.38 Scatter plot of the normalized longitudinal force versus the average slip ratio of front wheels during acceleration.

under those conditions to get consistent estimates of different slipslope values and hence accurate road friction coefficient estimation.

Simulation on Split-μ Road

To test other types of road surfaces and relate the estimated slope value to the road friction coefficients, a simulation scenario was designed in the CarSim environment. The vehicle initially travels on wet asphalt road and then its right wheels roll over snowy surface to mimic the unilaterally snow-covered condition that occurs in winter as shown in Figure 3.40. The magic tire models

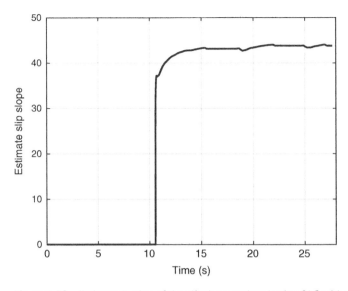

Figure 3.39 Estimated value of the slipslope updated using RLS with respect to time.

with longitudinal characteristics shown in Figure 3.31 for P225/50R17 tires were used during the simulation.

Since the vehicle's wheels from the left and right side may touch different surfaces during this test scenario, it is necessary to make separate estimations for individual wheels. During acceleration for the front-wheel-driven configuration, the vehicle longitudinal dynamics in (3.14), and the wheel rotational dynamics can be approximated as given below in terms of longitudinal forces of the front two tires, $F_{x,fl}$ and $F_{x,fr}$, as

$$m(\dot{v}_x + g \sin \theta) = F_{x,fl} + F_{x,fr} - F_{roll} - F_{air}, \tag{3.18}$$

$$I_w \dot{\omega}_{fl} = 0.5T_{dr} - R_{eff}F_{x,fl}, \tag{3.19}$$

$$I_w \dot{\omega}_{fr} = 0.5T_{dr} - R_{eff}F_{x,fr}, \tag{3.20}$$

Figure 3.40 Road surfaces designed in simulation environment.

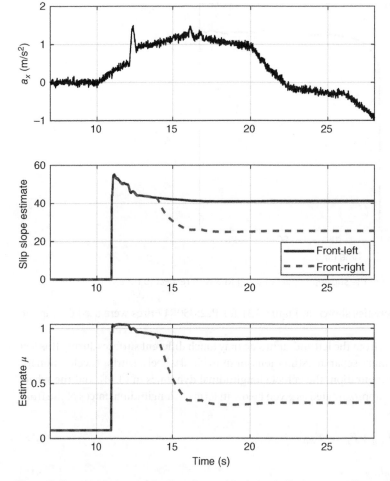

Figure 3.41 Vehicle longitudinal motion and estimated slipslope as well as road friction coefficient for individual wheels.

where the term $\dot{v}_x + g \sin \theta$ can be measured altogether by the accelerometer reading in the longitudinal direction, T_{dr} is the drive torque delivered to the front axle from the powertrain, R_{eff} is the average of the updating effective tire radius estimated for the rear two wheels in Section 3.6.1, and I_w is the moment of inertia for each wheel.

The derivatives of the wheel rotational speeds, $\dot{\omega}_{fl}$ and $\dot{\omega}_{fr}$, can be calculated using numerical differentiation. There are three unknowns left in Eqs. (3.18)–(3.20). It is hence possible to solve the longitudinal forces $F_{x,fl}$ and $F_{x,fr}$ for the individual wheels directly from the equation set. Note that it is also possible to estimate the individual tire longitudinal forces without using numerical differentiation by forming observers such as the Kalman filter based on Eqs. (3.18)–(3.20). The calculated individual longitudinal force is then related to the slip ratio of the corresponding tire to estimate the individual slipslope. The RLS with forgetting factor 0.999 at 100 Hz updating frequency is again applied with the corresponding estimation results shown in Figure 3.41 during acceleration.

It is observed that the estimation algorithm is triggered when enough acceleration is sensed. The slipslopes of both front tires show similar trend initially until the disparity appears around

14 seconds when the front-right tire rolls over from the wet asphalt to the snow-covered road surface. The road friction coefficients for both tires are also obtained through interpolation to the corresponding estimated slip slopes from the relationship shown in Figure 3.31. The disparity of the road surface of the left and right sides is evident from the estimation results. The road friction coefficient converges to around 0.9 and 0.32 for the left and right sides, respectively, which is reasonably close to the wet asphalt and snowy surface friction coefficients.

Note that the linear relationship between the tire longitudinal force and its slip ratio is an approximation as in Figure 3.31, where the slope of the force curve still varies slightly at small slip. Therefore, estimation error of the slipslope and correspondingly, road friction coefficient, is expected. Still, the slipslope method with RLS updating proves very effective and responsive from both the experimental and simulation results presented above.

3.7 Chapter Summary and Concluding Remarks

Background information useful for subsequent Chapters 4–7 of the book was introduced in this chapter. MIL and HIL simulation approaches and the experimental vehicles used were presented as they are all used in the evaluation of the path-tracking controllers and collision avoidance methods developed in subsequent Chapters 4–7. Unity and Unreal simulation environments and the LGSVL and CARLA simulators were presented for AV simulation along with methods of realistic simulation environment creation for the generation of raw AV sensor data. The slipslope method was introduced for tire-road friction estimation. Both experimental and simulated results showed that the slipslope method with RLS and forgetting factor result in accurate estimation of the tire-road friction.

References

1 S. Gelbal, S. Zhu, G. Anantharaman, B. Aksun Guvenc *et al.* (2019), "Cooperative collision avoidance in a connected vehicle environment," SAE Technical Paper 2019-01-0488, https://doi.org/10.4271/2019-01-0488.

2 M. R. Cantas, O. Kavas, S. Tamilarasan, S. Y. Gelbal, and L. Guvenc, "Use of hardware in the loop (HIL) simulation for developing connected autonomous vehicle (CAV) applications," SAE Technical Paper 0148-7191, 2019.

3 M. Ridvan Cantas, S. Fan, O. Kavas, S. Tamilarasan, L. Guvenc, S. Yoo, J. H. Lee, B. Lee, J. Ha. "Development of virtual fuel economy trend evaluation process," SAE Technical Paper 0148-7191, 2019.

4 Automated Driving Lab. (2018). "Realistic simulation of autonomous shuttle on OSU AV pilot test route". Available: https://www.youtube.com/watch?v=_yWiZWP0Rag&t=73s.

5 Ş. Y. Gelbal, S. Tamilarasan, M. R. Cantaş, L. Güvenç, and B. Aksun-Güvenç, "A connected and autonomous vehicle hardware-in-the-loop simulator for developing automated driving algorithms," in *IEEE International Conference on Systems, Man, and Cybernetics (SMC)*, 2017, pp. 3397–3402: IEEE.

6 D. J. Verburg, A. C. M. Van der Knaap, and J. Ploeg, "VEHIL: developing and testing intelligent vehicles," in *IEEE Intelligent Vehicle Symposium*, 2002, vol. 2, pp. 537–544: IEEE.

7 D. Zhao and H. Peng, "From the lab to the street: Solving the challenge of accelerating automated vehicle testing," *arXiv preprint arXiv:1707.04792*, 2017.

8 H. Liu and Y. Feng, "*Development of an Augmented Reality Environment for Connected and Automated Vehicle Testing*," University of Michigan, Ann Arbor, Transportation Research Institute, 2019.

9 L. Güvenç; I. M. C. Uygan; K. Kahraman; R. Karaahmetoglu; I. Altay; M. Sentürk; M. T. Emirler; A. E. H. Karci; B. A. Guvenc; E. Altug; M. C. Turan; Ö. S. Tas; E. Bozkurt; Ü. Ozguner; K. Redmill; A. Kurt; B. Efendioglu. "Cooperative adaptive cruise control implementation of team mekar at the grand cooperative driving challenge," *IEEE Transactions on Intelligent Transportation Systems*, vol. 13, no. 3, pp. 1062–1074, 2012.

10 M. T. Emirler, H. Wang, and B. A. Güvenç, "Socially acceptable collision avoidance system for vulnerable road users," *IFAC-PapersOnLine*, vol. 49, no. 3, pp. 436–441, 2016.

11 B. A. Guvenc, L. Guvenc, and S. Karaman, "Robust yaw stability controller design and hardware-in-the-loop testing for a road vehicle," *IEEE Transactions on Vehicular Technology*, vol. 58, no. 2, pp. 555–571, 2009.

12 E. Dinçmen, B. A. Güvenç, and T. Acarman, "Extremum-seeking control of ABS braking in road vehicles with lateral force improvement," *IEEE Transactions on Control Systems Technology*, vol. 22, no. 1, pp. 230–237, 2012.

13 M.T. Emirler, İ.M.C. Uygan, Ş.Y. Gelbal, M. Gözü, T.A. Böke, B. Aksun Güvenç, Güvenç, L., "Vehicle dynamics modelling and validation for hardware-in-the-loop testing of electronic stability control," *International Journal of Vehicle Design*, vol. 71, no. 1–4, pp. 191–211, 2016.

14 F. Coşkun, Ö. Tunçer, M. E. Karsligil, and L. Güvenç, "Real time lane detection and tracking system evaluated in a hardware-in-the-loop simulator," in *13th International IEEE Conference on Intelligent Transportation Systems*, 2010, pp. 1336–1343: IEEE.

15 O. U. Acar, L. Güvenç, and E. Altuğ, "Hardware-in-the-loop testing of automatic lift dropping system for heavy trucks," *Journal of Intelligent and Robotic Systems* 98, pp. 1–11, 2019.

16 M. R. Cantas and L. Guvenc, "Camera based automated lane keeping application complemented by GPS localization based path following," SAE Technical Paper 0148-7191, 2018.

17 K. M. Cime, M. R. Cantas, G. Dowd, L. Guvenc, B. A. Guvenc, A. Mittal, A. Joshi, J. Fishelson. "Hardware-in-the-loop, traffic-in-the-loop and software-in-the-loop autonomous vehicle simulation for mobility studies," SAE Technical Paper 0148-7191, 2020.

18 PTV Group Vissim. PTV Vissim. Available: https://company.ptvgroup.com/en-us/.

19 Mathworks, "MATLAB Automated Driving Toolbox". Available: https://www.mathworks.com/products/automated-driving.html.

20 Mathworks, "MATLAB ROS Toolbox". Available: https://www.mathworks.com/products/ros.html.

21 Mathworks, "MATLAB Simscape Driveline". Available: https://www.mathworks.com/products/simscape-driveline.html.

22 X. Li, A. C. A. Doss, B. A. Guvenc, and L. Guvenc, "Pre-deployment testing of low speed, urban road autonomous driving in a simulated environment," *SAE International Journal of Advances and Current Practices in Mobility*, vol. 2, no. 2020-01-0706, pp. 3301–3311, 2020.

23 L. Guvenc, B. Aksun-Guvenc, X. Li, A. C. A. Doss, K. Meneses-Cime, and S. Y. Gelbal, "Simulation environment for safety assessment of CEAV deployment in Linden," *arXiv preprint arXiv:2012.10498*, 2020.

24 Mathworks, "MATLAB RoadRunner". Available: https://www.mathworks.com/products/roadrunner.html.

25 LG Electronics, "SVL SIMULATOR". Available: https://www.svlsimulator.com/.

26 P. Cignoni, M. Callieri, M. Corsini, M. Dellepiane, F. Ganovelli, and G. Ranzuglia, "Meshlab: an open-source mesh processing tool," in *Eurographics Italian Chapter Conference*, 2008, vol. 2008, pp. 129–136: Salerno, Italy.

27 Unity, "EasyRoads 3D". Available: https://www.easyroads3d.com/

28 T. A. Foundation, "Autoware.auto". Available: https://www.autoware.auto/

29 A. Dosovitskiy, G. Ros, F. Codevilla, A. Lopez, and V. Koltun, "CARLA: an open urban driving simulator," in *Conference on Robot Learning*, 2017, pp. 1–16: PMLR.

30 Google, "Google Earth Studio". Available: https://www.google.com/earth/studio/

31 B. Wu, A. Wan, X. Yue, and K. Keutzer, "Squeezeseg: convolutional neural nets with recurrent CRF for real-time road-object segmentation from 3D lidar point cloud," in *IEEE International Conference on Robotics and Automation (ICRA)*, 2018, pp. 1887–1893: IEEE.

32 Anonymous, "Blender Foundation". Available: https://www.blender.org/

33 R. B. Rusu and S. Cousins, "3D is here: point cloud library (PCL)," in *IEEE International Conference on Robotics and Automation*, 2011, pp. 1–4: IEEE.

34 AliceVision, "Meshroom". Available: https://github.com/alicevision/meshroom.

35 X. Li, S. Zhu, S. Y. Gelbal, M. R. Cantas, B. A. Guvenc, and L. Guvenc, "A unified, scalable and replicable approach to development, implementation and HIL evaluation of autonomous shuttles for use in a smart city," SAE Technical Paper 0148-7191, 2019.

36 B. Wen, S. Y. Gelbal, B. A. Guvenc, and L. Guvenc, "Localization and perception for control and decision-making of a low-speed autonomous shuttle in a campus pilot deployment," *SAE International Journal of Connected and Automated Vehicles*, vol. 1, no. 12-01-02-0003, pp. 53–66, 2018.

37 J. Ackermann, "*Robust Control: The Parameter Space Approach*". Springer Science & Business Media, 2012.

38 S. Y. Gelbal, N. Chandramouli, H. Wang, B. Aksun-Guvenc, and L. Guvenc, "A unified architecture for scalable and replicable autonomous shuttles in a smart city," in *IEEE International Conference on Systems, Man, and Cybernetics (SMC)*, 2017, pp. 3391–3396: IEEE.

39 S. Zhu, S. Y. Gelbal, X. Li, M. R. Cantas, B. Aksun-Guvenc, and L. Guvenc, "Parameter Space and Model Regulation Based Robust, Scalable and Replicable Lateral Control Design for Autonomous Vehicles," *57th IEEE Conference on Decision and Control*, Miami Beach, Florida, 2018, pp. 6963–6969.

40 C. K. Song, M. Uchanski, and J. K. Hedrick, "Vehicle speed estimation using accelerometer and wheel speed measurements," SAE Technical Paper 0148-7191, 2002.

41 K. J. Åström and B. Wittenmark, "*Adaptive Control*". Courier Corporation, 2013.

42 D. Fischer, S. Stoelzl, and A. Koebe, "Friction coefficient estimation from camera and wheel speed data," ed: Google Patents, 2017.

43 G. Erdogan, L. Alexander, and R. Rajamani, "Estimation of tire-road friction coefficient using a novel wireless piezoelectric tire sensor," *IEEE Sensors Journal*, vol. 11, no. 2, pp. 267–279, 2010.

44 A. Miyoshi, "Pneumatic tire with specifically arranged strain sensors," ed: Google Patents, 2009.

45 A. Pohl, R. Steindl, and L. Reindl, "The "intelligent tire" utilizing passive SAW sensors measurement of tire friction," *IEEE Transactions on Instrumentation and Measurement*, vol. 48, no. 6, pp. 1041–1046, 1999.

4

Path Description and Generation

This chapter starts with a review of some of the commonly used path modeling and generation methods available in the literature. The specific path generation methods presented in this chapter are the use of clothoids, Bezier curves, and polynomial splines. Re-distribution of waypoints based on route curvature for a route obtained by data collection or by direct extraction from a map is introduced first, followed by smoothing of the waypoints data to reduce the scatter inherent in such experimentally collected values. Clothoids, Bezier curves, and polynomial splines are introduced and discussed in detail next. The least squares curve fit algorithm for polynomial spline coefficient determination for a path divided into segments is presented. The chapter ends with algorithms for computation of path-tracking error after the desired path is determined in the form of discrete waypoints or in the form of polynomial splines.

4.1 Introduction

This chapter introduces different path descriptions including discrete-waypoint and polynomial-spline representations. The raw data points for path reference could be from existing roadmaps or recorded positions with GPS sensor. These raw data points are often noisy or too scattered and require further processing to form a smooth path description. We compared both the discrete and continuous path description in the form of discrete waypoints and polynomial splines with corresponding path smoothing and fitting techniques, respectively. The tracking errors of current vehicle position and postures from the reference path are necessary as the feedback for an accurate path following control. The calculation of tracking errors is detailed for both discrete waypoints and spline path representations.

4.2 Discrete Waypoint Representation

One straightforward way of creating a desired path for an autonomous vehicle is to drive the vehicle on the intended path while periodically recording its position. A high-accuracy GPS unit is usually used in this approach to record absolute position. This sequence of positions becomes a list of waypoints representing the path. This is the simplest way of creating a path representation and suffers from inaccuracies and errors inherent in the sensors used, the GPS sensor in this case. Even if the vehicle follows a straight path, the recorded sequence of GPS waypoints may not lie on this line exactly but would instead scatter around it. There may also be an unnecessarily high number of points being recorded especially in straight portions of the desired path. The sequence

Autonomous Road Vehicle Path Planning and Tracking Control, First Edition.
Levent Güvenç, Bilin Aksun-Güvenç, Sheng Zhu, and Şükrü Yaren Gelbal.

of waypoints recorded while driving along the desired path are, therefore, not used directly to avoid zigzag type driving behavior and for computational efficiency. They have to be smoothed to avoid the zigzag-type scatter and reduced in number where too many waypoints are not needed. The reduction in the number of waypoints, i.e. waypoint density redistribution, and smoothing of the waypoints are presented next in the form of two successive steps.

Step 1: The waypoint density needs to be redistributed along the path. For a straight road segment, just a few discrete waypoints are enough to fully describe it. On the other hand, more waypoints are needed at corners where large curvature occurs. Thus, the path waypoints are redistributed with a density in proportion to the road curvature κ, which is calculated from the original waypoints without smoothing using

$$\kappa_i = \frac{\nabla x_i \nabla^2 y_i - \nabla^2 x_i \nabla y_i}{\left((\nabla x_i)^2 + (\nabla y_i)^2\right)^{3/2}}, \tag{4.1}$$

where ∇x_i, ∇y_i, $\nabla^2 x_i$, and $\nabla^2 y_i$ are the first-order and second-order numerical gradients of the ith waypoint's coordinates. Since the waypoint data has scatter, the curvature values obtained from them using Eq. (4.1) will also have scatter that needs to be smoothed. The curvature values are smoothed with a Savitzky–Golay filter [1]. The Savitzky–Golay filter is a digital filter used for smoothing discrete data using convolution. Successive subsets of adjacent data points are fit linear least squares low-order polynomials. When the data points are equally spaced, an analytical solution to the least-squares equations is found as a single set of convolution coefficients. The Matlab® command *sgolayfilt* can be used to apply the Savitzky–Golay filter.
The waypoints in the original data that are retained are chosen to satisfy

$$\Delta s_i = s_{i+1} - s_i \le \frac{0.1}{\max_{s_i < s < s_{i+1}} \{|\kappa(s)|\}} \tag{4.2}$$

between successive retained waypoints where s_i is the station at the ith waypoint, Δs_i is the distance between neighboring waypoints and is limited to be within the range of $[1,16]$ m. These retained waypoints after adjustment are labeled as "adjusted" in Figure 4.1, with higher densities observed at large curvature segments. Note that the adjusted waypoints still have scatter as they are a subset of the original, larger waypoint data.

Step 2: The coordinates of the adjusted waypoints $\mathbf{p}_1, \ldots, \mathbf{p}_N$ obtained during the above step are further modified through least square optimization to improve smoothness. The optimized waypoints $\mathbf{q}_1, \ldots, \mathbf{q}_N$ are found by solving the following additive function:

$$\underset{\mathbf{q}_1, k, \mathbf{q}_N}{\arg\min} \left\{ \sum_{i=1}^{N} \|\mathbf{q}_i - \mathbf{p}_i\|^2 - w \sum_{i=2}^{N-1} \frac{(\mathbf{q}_{i+1} - \mathbf{q}_i) \cdot (\mathbf{q}_i - \mathbf{q}_{i-1})}{\|\mathbf{q}_{i+1} - \mathbf{q}_i\| \|\mathbf{q}_i - \mathbf{q}_{i-1}\|} \right\}, \tag{4.3}$$

where the first summation is the accumulated deviation of $\mathbf{q}_i$ from $\mathbf{p}_i$. The latter summation represents the cosine value of the angle between the neighboring waypoint segments with a weight w, which penalizes the heading inconsistency along the path. The overall waypoint adjustment procedure is shown in Table 4.1.
We take a cornering road segment to demonstrate the waypoints adjustment process. In this example, the vehicle poses including its position and heading when driving along the path are recorded as the raw discrete waypoints from the virtual GPS sensor in an autonomous-driving simulator. Figure 4.1a shows comparisons of waypoint positions before processing, after density adjustment, and after optimization. Notice that after adjustment, the waypoints redistribute

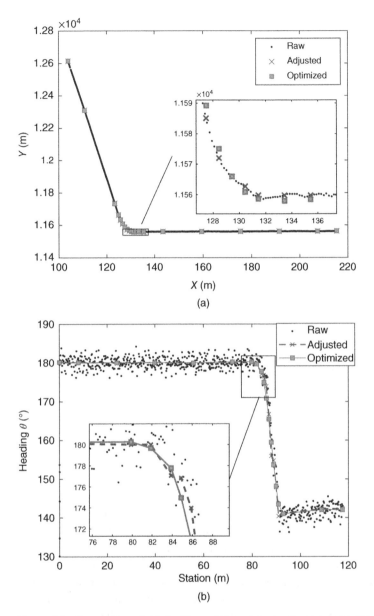

Figure 4.1 Comparison of: (a) positions of the original, adjusted, and optimized waypoints, as well as (b) their headings along the station.

themselves according to the road curvature, i.e. denser waypoints around the corners, which significantly reduces the number of waypoints required to describe the path. After optimization, the optimized waypoints (in squares) further improve the path smoothness as compared to the other two cases in rugged shape, as shown in the zoomed subplot. Figure 4.1b further shows the comparisons of heading angles for the three cases, which again reveals the improvement of heading consistency after waypoint adjustment and optimization.

Table 4.1 Discrete waypoint processing procedures.

Algorithm: Waypoint Adjustment

1. Given raw waypoints $\mathbf{p} = \{\mathbf{p}_1, \mathbf{p}_2, \ldots, \mathbf{p}_i, \ldots, \mathbf{p}_n\}$
$$\mathbf{p}_i = \{x_i, y_i\}$$
2. Waypoint density adjustment:
   a. (optional) Interpolate the original waypoints for distributing them evenly.
   b. Update curvature values with equation (4.1) and smooth them using a Savitzky-Golay filter.
   c. Apply greedy algorithm starting from one end to determine positions of new waypoints such that the condition in equation (4.2) is satisfied:

Pseudo code:

```
s_new = [0]; Δs = 16;
for the ith waypoints {
    if (Δs < s_i − s_new [end]) {
        s_new.append( s_new[end] + Δs );
        Δs = 16;                    // reset
    }
    Δs = min{ Δs, 0.1/abs(κ_i(s)) };
    Δs = (Δs < 1) ? 1 : Δs; // lower limit
}
x_new = interpolation( {s_i}, {x_i}, s_new );
y_new = interpolation( {s_i}, {y_i}, s_new );
```

---

3. Optimize waypoints to improve heading consistency by solving (4.3).

4.3 Parametric Path Description

Other than the discrete waypoint path description of Section 4.2, there are also methods describing the path in an analytical or parametric form. The previously introduced discrete waypoint description has the benefits of easy formation, but the transition of positions and headings in-between two discrete neighboring waypoints might not be favorably smooth and would require either interpolation or fitting techniques. The parametric path description, on the other hand, has the advantage of path continuity and smoothness in term of its trace, heading, and even curvature. One might ask, if the parametric path description is so superior, why would we all not favor it over the discrete waypoint representation? The headache of the parametric path description lies in the process of choosing the proper parametric form such as order or segmentation and to decide the appropriate numerical parameter values. Overfitting also becomes an issue later when exceedingly high polynomial order is used for splines, for example.

The general idea of parametric path description is demonstrated in Figure 4.2, where λ is a dimensionless distance parameter along the path. The coordinate of any point on the path is described generically as follows:

$$X_P = X_P(\lambda), \tag{4.4}$$

$$Y_P = Y_P(\lambda). \tag{4.5}$$

Figure 4.2 Illustration of parametric path description.

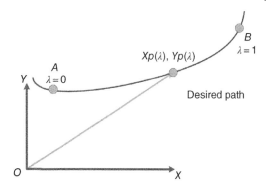

There are a lot of different functions that can be used on the right-hand side of Eqs. (4.4) and (4.5) to describe the desired path expressed in terms of λ. Out of these possible functions, clothoids and Bezier curves are presented next. This is followed by a detailed description of polynomial splines as they are used in Chapter 5 also.

4.3.1 Clothoids

A curve can in general be described by the following differential equations:

$$x'(s) = \cos\theta(s), \quad x(0) = x_0, \tag{4.6}$$

$$y'(s) = \sin\theta(s), \quad y(0) = y_0, \tag{4.7}$$

$$\theta'(s) = \kappa(s), \quad \theta(0) = \theta_0, \tag{4.8}$$

where s is the arc length of the curve, $x'(s)$, $y'(s)$, and $\theta'(s)$ are the derivatives of coordinates and heading with respect to the arc length s. Note that the parameter λ becomes the distance along the curve (arc length) s in this description. When the curvature changes linearly in terms of the arc length, i.e. when (see [2])

$$\kappa(s) := \kappa_0 + \kappa's, \tag{4.9}$$

the curve is called a clothoid. In Eq. (4.9), κ_0 is the initial curvature at $s = 0$ m, and the constant coefficient κ' is the rate of curvature change with respect to the arc length s. This curve traces out a trajectory in the x–y–s space known as the Cornu spiral [3] which is shown as the solid curve in Figure 4.3. The dashed curves in Figure 4.3 are the projection of the Cornu spiral onto the x–s, y–s, and y–x planes.

The representations in Eqs. (4.6)–(4.9) can be converted to the more general parametric form of a clothoid spiral curve as (see [2]):

$$x(s) = x_0 + \int_0^s \cos\left(\theta_0 + \kappa_0\lambda + \frac{1}{2}\kappa'\lambda^2\right) d\lambda, \tag{4.10}$$

$$y(s) = y_0 + \int_0^s \sin\left(\theta_0 + \kappa_0\lambda + \frac{1}{2}\kappa'\lambda^2\right) d\lambda. \tag{4.11}$$

Clothoid curves are commonly applied in real-world applications such as highway and railroad design, map projection, typography, and digital vector drawing [4–6] due to their inherently smooth, linear, and consistent transition of curvature change.

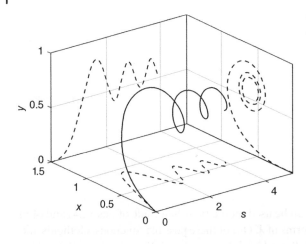

Figure 4.3 Cornu spiral in the $x-y-s$ space and its plane projections, at $\kappa_0 = -0.05$ and $\kappa' = 3\pi/4$.

Clothoids are also favored by researchers of self-driving path planning due to these features as well as their simplicity. The linearly changing curvature of the clothoid enables the autonomous vehicle to ride smoothly by turning its steering wheel at a constant rate and hence being beneficial to passenger comfort. A single clothoid curve has been applied in very specific cases such as lane changing and emergency collision avoidance [7, 8]. The authors of reference [9] managed to optimize the racing path for their autonomous racing car by representing the turning corner by two clothoids that are interconnected with a circular arc. A more feasible solution to equip the vehicle with the reactive capability for collision avoidance can be seen in [10], where the path planning is selected from clothoid tentacles. In the approach of [10], the clothoid tentacles are generated by sampling the curvature rate κ'. The initial curvature κ_0 is determined by the steering angle δ_0 from the vehicle kinematics as

$$\kappa_0 = \frac{\tan \delta_0}{L}, \tag{4.12}$$

where L is the vehicle wheelbase. The length of each clothoid tentacle is related to current vehicle speed v_x as

$$s_f = \begin{cases} 7v_x - 5, & v_x > 1 \, \text{m/s} \\ 2, & v_x \leq 1 \, \text{m/s} \end{cases}. \tag{4.13}$$

The curvature along individual clothoids is not allowed to exceed the maximum allowed value κ_{max} that is limited by the maximum lateral acceleration $a_{y,max}$ as

$$\kappa_{max} = \frac{a_{y,max}}{v_x^2}. \tag{4.14}$$

The terminal curvature of the leftmost and the rightmost tentacle are set to reach this maximum allowed κ_{max} as shown below:

$$\kappa_0 + \kappa'_{leftmost} s_f = \kappa_{max}, \tag{4.15}$$

$$\kappa_0 - \kappa'_{rightmost} s_f = -\kappa_{max}. \tag{4.16}$$

Therefore, the rate of curvature κ' could be sampled within the range of

$$\frac{-\kappa_{max} - \kappa_0}{s_f} \leq \kappa' \leq \frac{\kappa_{max} - \kappa_0}{s_f}. \tag{4.17}$$

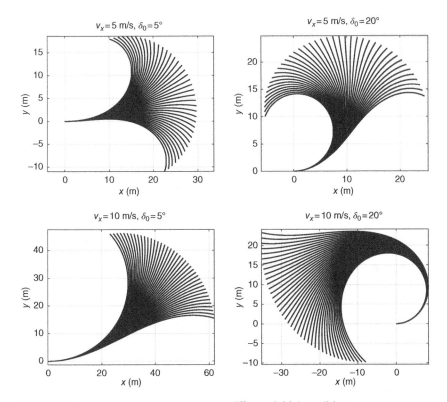

Figure 4.4 Clothoid tentacles generated at different initial condition.

Figure 4.4 shows the generated clothoid tentacles that are considered feasible at different speeds and initial steering angle values. At every planning cycle, these tentacles will be generated and one of them will then be selected based on the cost function along with the safety of other road users.

Of course, with only three variable parameters of (κ_0, κ', and s), the clothoid may not be able to satisfy a large number of arbitrary conditions including initial curvature, terminal curvature, heading constraints, and terminal position. Given the need for additional parameters to satisfy these constraints, Kelly resolved this by increasing the number of higher-order terms to the curvature polynomial (4.9) so that it became [3]

$$\kappa(s) := a + bs + cs^2 + ds^3 + L, \tag{4.18}$$

of which the polynomial coefficients could then be determined by optimization techniques. One can also resolve this issue by seeking more complex parametric curve representations that inherently have a higher number of variable parameters.

4.3.2 Bezier Curves

Bezier curves are parametric curves that are used frequently in computer graphics to generate smooth curves that can be scaled indefinitely. A Bezier curve is shaped by its control points. For example, a quadratic Bezier curve is the path traced by the function $\mathbf{B}(\lambda)$, given control points $\mathbf{P}_0$, $\mathbf{P}_1$, and $\mathbf{P}_2$ and satisfying

$$\mathbf{B}(\lambda) = (1 - \lambda)[(1 - \lambda)\mathbf{P}_0 + \lambda\mathbf{P}_1] + \lambda[(1 - \lambda)\mathbf{P}_1 + \lambda\mathbf{P}_2], \quad 0 \le \lambda \le 1. \tag{4.19}$$

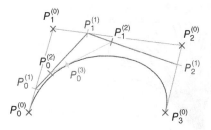

Figure 4.5 The forming of Bezier curve with De Casteljau's algorithm [11].

The above form can be interpreted as the linear interpolation of corresponding points on the linear Bezier curves from $\mathbf{P}_0$ to $\mathbf{P}_1$, and from $\mathbf{P}_1$ to $\mathbf{P}_2$. Rearrangement of expression (4.19) becomes

$$\mathbf{B}(\lambda) = (1 - \lambda)^2 \mathbf{P}_0 + 2(1 - \lambda)\lambda \mathbf{P}_1 + \lambda^2 \mathbf{P}_2, \quad 0 \leq \lambda \leq 1. \tag{4.20}$$

The first and second derivative of the Bezier curve is continuous as

$$\mathbf{B}'(\lambda) = 2(1 - \lambda)(\mathbf{P}_1 - \mathbf{P}_0) + 2\lambda(\mathbf{P}_2 - \mathbf{P}_1), \tag{4.21}$$

$$\mathbf{B}''(\lambda) = 2(\mathbf{P}_2 - 2\mathbf{P}_1 + \mathbf{P}_0). \tag{4.22}$$

Similarly, Bezier curves at higher orders can be derived as illustrated in Figure 4.5. The recursive process to form the Bezier curve is named the De Casteljau's algorithm [11]. One feature of Bezier curves is that the tangents at its starting and ending point are toward the next and next-to-last control points, respectively.

The curvature of the Bezier curve is also continuous. The curvature at point $\mathbf{B}(x(\lambda), y(\lambda))$ can be analytically derived by substitution of Eqs. (4.21) and (4.22) into

$$\kappa(\lambda) = \frac{x'(\lambda)y''(\lambda) - x''(\lambda)y'(\lambda)}{(x'^2(\lambda) + y'^2(\lambda))^{3/2}}. \tag{4.23}$$

It is seen that the Bezier curve is determined once the locations of its control points are fixed. So the Bezier curve is designed by properly choosing its control points. Selection from geometric relationships with surrounding objects or traversable corridors is one way. Jolly et al. [12] relates the two intermediate control points to the lengths of control point connections, and iteratively find the proper lengths satisfying obstacle avoidance for robots. References [13, 14] utilize the geometry of the corridor at corners to decide locations of control points and eventually find one Bezier curve that satisfies the corridor constraints. Optimization techniques are also seen in determining the appropriate control points to form a smooth Bezier path that satisfies corridor constraints [15]. Optimization is also used when fitting a coarse path with the Bezier curve, such as in the automatic parking case in [16], and in [17] where the Bezier curve is used to fit piecewise-linear segments via optimization.

An example of Bezier curve application in path generation at a corner is given below as shown in Figure 4.6. The samples of Bezier curves are generated by sampling intermediate control points according to the geometry similar to [14]. The square points in the plot represent the locations of the five control points for one curve. Then, a suitable curve (Figure 4.6) which is inside the corridor and has the lowest cost in terms of a defined criterion like minimizing vicinity to the centerline can be selected out of these samples. While the example given is for an indoor application, the same approach is applicable to outdoor, on-road applications when the corridor boundaries are replaced by lanes or the boundaries of obstacles on the road like other vehicles and road boundaries like the curb.

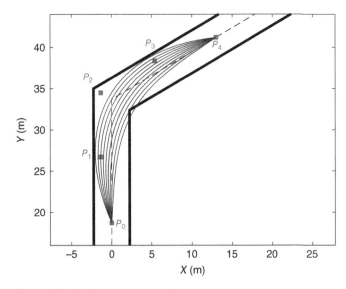

Figure 4.6 Bezier curve generation at a corner.

4.3.3 Polynomial Spline Description

A spline consists of continuous segments represented by polynomials with the coordinates of the ith segment described by

$$x_i(\lambda) = \sum_{k=0}^{p} a_{i,k} \lambda^k, \tag{4.24}$$

$$y_i(\lambda) = \sum_{k=0}^{p} b_{i,k} \lambda^k, \tag{4.25}$$

where p is the chosen polynomial order, parameter $\lambda \in [0,1]$, $a_{i,k}$ and $b_{i,k}$ are the polynomial coefficients for the kth order item of the ith segment.

Polynomial curve fitting is used to obtain the coefficients of best fit to the collected or recorded waypoints forming the path. The fitting process for coefficients $a_{i,k}$ and $b_{i,k}$ is conducted separately but in a similar manner. We use the fitting process of coefficients $a_{i,k}$ for demonstration. The x coordinate of any point at the ith segment in (4.24) can be rearranged to the regression form:

$$x_i = \varphi^T \theta_i, \tag{4.26}$$

with the regressor $\varphi = [1, \lambda^1, \ldots, \lambda^p]^T$ and the parameter to be estimated $\theta_i = [a_{i,0}, a_{i,1}, \ldots, a_{i,p}]^T$.

The n waypoints in total are divided into m segments with the segment nodes being either hand-picked or evenly distributed. Denote the jth waypoint in the ith segment with the subscript "i, j" with its coordinates denoted as $(x_{i,j}, y_{i,j})$. The independent variable value $\lambda_{i,j}$ for this waypoint is also assigned to be between 0 and 1 according to its relative distance within this segment. Rearrange the x coordinates and the corresponding regressors of these waypoints as well as the polynomial parameters to be estimated into vector/matrix form as

$$X = [\underbrace{x_{1,1}, x_{1,2}, \ldots, x_{1,n_1}}_{\text{1st segment}}, \underbrace{x_{2,1}, x_{2,2}, \ldots, x_{2,n_2}}_{\text{2nd segment}}, \ldots, \underbrace{x_{m,1}, x_{m,2}, \ldots, x_{m,n_m}}_{\text{mth segment}}]^T \in \mathbb{R}^n, \tag{4.27}$$

$$\Phi = \begin{bmatrix} \varphi^T(\lambda_{1,1}) & 0 & \cdots & & & 0 \\ \varphi^T(\lambda_{1,2}) & & & & & \\ \vdots & & & & & \\ \varphi^T(\lambda_{1,n_1}) & & & & & \\ & \varphi^T(\lambda_{2,1}) & & & & \\ & \varphi^T(\lambda_{2,2}) & & & & \\ & \vdots & & & & \\ 0 & \varphi^T(\lambda_{2,n_2}) & & & & \\ \vdots & & \ddots & & & \\ & & & \varphi^T(\lambda_{m,1}) & & \\ & & & \varphi^T(\lambda_{m,2}) & & \\ & & & \vdots & & \\ 0 & & & \varphi^T(\lambda_{m,n_m}) & & \end{bmatrix} \in \mathbb{R}^{n \times (p+1)m}, \tag{4.28}$$

$$\Theta = [\theta_1^T, \theta_2^T, \dots, \theta_m^T]^T \in \mathbb{R}^{(p+1)m}, \tag{4.29}$$

where n_i is the number of waypoints in the ith segment. After introducing these matrix notations, the polynomial fitting problem can be neatly described below to find the parameters Θ that minimize the sum of squared residuals:

$$\min_{\Theta} \|\Phi\Theta - X\|^2. \tag{4.30}$$

The parameters Θ found in this way would not be satisfactory since the connection and continuity between segments are not ensured. Therefore, continuity constraints up to order q $(q < p)$ are defined between neighboring segments as follows:

$$x_i(1) = x_{i+1}(0), \tag{4.31}$$

$$\frac{dx_i}{d\lambda}(1) = \frac{dx_{i+1}}{d\lambda}(0), \tag{4.32}$$

$$\vdots$$

$$\frac{d^q x_i}{d\lambda^q}(1) = \frac{d^q x_{i+1}}{d\lambda^q}(0). \tag{4.33}$$

They can be expanded into the following algebraic equations:

$$(a_{i,0} + a_{i,1} + \cdots + a_{i,p}) - a_{i+1,0} = 0, \tag{4.34}$$

$$(a_{i,1} + 2a_{i,2} + \cdots + pa_{i,p}) - a_{i+1,1} = 0, \tag{4.35}$$

$$\vdots$$

$$\left(q!a_{i,q} + (q+1)!a_{i,q+1} + \cdots + \frac{p!}{(p-q)!}a_{i,p} \right) - q!a_{i+1,q} = 0, \tag{4.36}$$

and further, more neatly to the below matrix representations:

$$A_i\theta_i + B_i\theta_{i+1} = 0, \tag{4.37}$$

where the matrices are given by

$$A_i = \begin{bmatrix} 1 & 1 & 1 & \cdots & 1 & \cdots & 1 \\ 0 & 1 & 2 & \cdots & q & \cdots & p \\ \vdots & \vdots & \vdots & & \vdots & & \vdots \\ 0 & 0 & 0 & \cdots & q! & \cdots & \frac{p!}{(p-q)!} \end{bmatrix} \in \mathbb{R}^{(q+1)\times(p+1)}, \tag{4.38}$$

$$B_i = \begin{bmatrix} -1 & 0 & \cdots & 0 & \cdots & 0 \\ 0 & -1 & \cdots & 0 & \cdots & 0 \\ \vdots & \vdots & \ddots & 0 & & 0 \\ 0 & 0 & 0 & -1 & \cdots & 0 \end{bmatrix} \in \mathbb{R}^{(q+1)\times(p+1)}. \tag{4.39}$$

Similarly we can obtain the qth-order continuity constraints between other neighboring segments. These algebraic equations can then be re-organized to an equality constraint in the vector form:

$$\Gamma\Theta = 0, \tag{4.40}$$

where matrix Γ can be represented as:

$$\Gamma = \begin{bmatrix} A_1 & B_1 & 0 & \cdots & 0 & 0 \\ 0 & A_2 & B_2 & \cdots & 0 & 0 \\ \vdots & \vdots & \vdots & \ddots & \vdots & \vdots \\ 0 & 0 & 0 & \cdots & A_{m-1} & B_{m-1} \end{bmatrix} \in \mathbb{R}^{(q+1)(m-1)\times(p+1)m}, \tag{4.41}$$

or

$$\Gamma = \begin{bmatrix} A_1 & B_1 & 0 & \cdots & 0 & 0 \\ 0 & A_2 & B_2 & \cdots & 0 & 0 \\ \vdots & \vdots & \vdots & \ddots & \vdots & \vdots \\ 0 & 0 & 0 & \cdots & A_{m-1} & B_{m-1} \\ B_m & 0 & 0 & \cdots & 0 & A_m \end{bmatrix} \in \mathbb{R}^{(q+1)m\times(p+1)m}, \tag{4.42}$$

if all the m segments form a closed loop.

The minimizing function (4.30) and the equality constraint (4.40) taken together form a constrained least square problem. This problem can be solved by first defining the Lagrangian function with the introduced Lagrangian multipliers $\gamma \in \mathbb{R}^{(q+1)}$:

$$\mathcal{L}(\Theta, \gamma) = \|\Phi\Theta - X\|^2 + \gamma^T\Gamma\Theta = (\Phi\Theta - X)^T(\Phi\Theta - X) + \gamma^T\Gamma\Theta. \tag{4.43}$$

The optimum $\hat{\Theta}$ satisfies the first-order necessary conditions as follows:

$$\frac{\partial\mathcal{L}}{\partial\Theta}(\Theta, \gamma) = 2\Phi^T(\Phi\Theta - X) + \Gamma^T\gamma = 0, \tag{4.44}$$

$$\frac{\partial\mathcal{L}}{\partial\gamma}(\Theta, \gamma) = \Gamma\Theta = 0, \tag{4.45}$$

where Eq. (4.45) is the same as the equality constraint (4.40). The above conditions are organized together below as the Karush–Kuhn–Tucker (KKT) or first derivative conditions given by

$$\begin{bmatrix} 2\Phi^T\Phi & \Gamma^T \\ \Gamma & 0 \end{bmatrix} \begin{bmatrix} \hat{\Theta} \\ \gamma \end{bmatrix} = \begin{bmatrix} 2\Phi^TX \\ 0 \end{bmatrix}. \tag{4.46}$$

The left-hand side of the KKT matrix above is invertible if and only if Γ has independent rows and $\begin{bmatrix} \Phi \\ \Gamma \end{bmatrix}$ has independent columns. This implies that $q + 1 \le (p + 1)m$ and $n + q + 1 \ge (p + 1)m$, the former of which indicates that the continuity order should not be greater than the total number of polynomial parameters. The latter indicates that the number of waypoints should be reasonably large, at least higher than the total number of polynomial parameters. Once the polynomial orders, number of polynomial segments, and continuity orders are appropriately selected to satisfy these conditions, the nonsingularity of the KKT matrix is ensured and the parameters to be estimated can be found uniquely by the inversion:

$$\begin{bmatrix} \hat{\Theta} \\ \gamma \end{bmatrix} = \begin{bmatrix} 2\Phi^T\Phi & \Gamma^T \\ \Gamma & 0 \end{bmatrix}^{-1} \begin{bmatrix} 2\Phi^T X \\ 0 \end{bmatrix}. \tag{4.47}$$

Figure 4.7 shows fitting results of discrete raw waypoints by splines with different polynomial and continuity orders. The segment nodes can be picked manually or distributed according to curvature similar to the waypoint density adjustment presented in Section 4.3. In this example, the segment nodes are evenly distributed and shown with diamond shapes in Figure 4.7a. It is observed all three splines fit the raw data point coordinates quite well.

One benefit of the spline path description is that not only are the path coordinates continuous but also the path heading and curvature, as shown in Figure 4.7b,c, which is beneficial for the path following control as this representation provides a smooth reference path to follow. In Figure 4.7, we also made comparisons of spline fitting between polynomial orders p and continuity orders q, with the combinations of (p,q) chosen as (4,2), (4,1), (8,2), respectively. It is noticed that the first-order continuity is not favored as the curvature change is not continuous between polynomial segments (Figure 4.7c) which would result in sudden change of steering for path following. There is also risk of overfitting if the polynomial order is chosen too high as is observed for the eighth-order spline case in Figure 4.7c, where curvature tends to be like the raw waypoints and, hence, oscillatory.

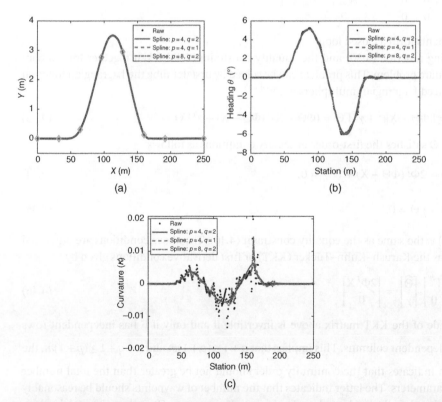

Figure 4.7 Positions (a), headings (b), and curvatures (c) of the original waypoints and fitted splines at different orders of polynomials and continuity.

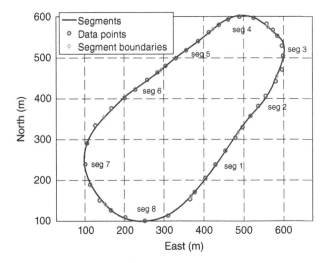

Figure 4.8 Data points of a track and its fitted result with polynomial splines [18].

Another polynomial spline example is given in Figure 4.8, where the sparsely distributed discrete data points of a track are fitted with the constrained least-square fitting method presented above. It can be seen that the generated path is smooth and remains close to the original data points.

4.4 Tracking Error Calculation

Tracking error refers to the deviations of the vehicle position and posture from the reference path. A quantitative measure of tracking error is needed by the path-following controller in order to compute the required steering corrections and is explained in this section. Please note that the details involved in computing tracking error may be slightly different depending on how the path is described.

4.4.1 Tracking Error Computation for a Discrete Waypoint Path Representation

Section 4.2 covers the path representation in discrete waypoints and the smoothening optimization procedure. It is then necessary to calculate the tracking error of the vehicle posture located at P with respect to the path described by discrete waypoints, $\{q_k\}$, $k = 1,2, \ldots$, as illustrated in Figure 4.9a. The first step naturally is to find the nearest waypoint to the vehicle position. Then, the segment of

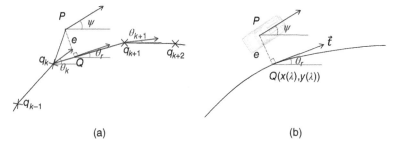

(a)　　　　　　　　　　(b)

Figure 4.9 Illustration of tracking errors for (a) discrete waypoint path description, and (b) spline path.

interest with the nearest waypoint at one end is found as the segment $\overrightarrow{q_k q_{k+1}}$ as shown in Figure 4.9a and satisfies

$$(\overrightarrow{q_k P} \cdot \overrightarrow{q_k q_{k+1}})(\overrightarrow{q_{k+1} P} \cdot \overrightarrow{q_k q_{k+1}}) \leq 0. \tag{4.48}$$

The closest point along the path to the vehicle position P is then approximated at point Q in Figure 4.9a, where $\overrightarrow{PQ}$ is perpendicular to $\overrightarrow{q_k q_{k+1}}$. The ratio of the projection of $\overrightarrow{q_k P}$ on $\overrightarrow{q_k q_{k+1}}$ to the length of $\overrightarrow{q_k q_{k+1}}$ is

$$\chi = \frac{\left|\overrightarrow{q_k Q}\right|}{\left|\overrightarrow{q_k q_{k+1}}\right|} = \frac{\overrightarrow{q_k P} \cdot \overrightarrow{q_k q_{k+1}}}{\left|\overrightarrow{q_k q_{k+1}}\right|^2}. \tag{4.49}$$

The reference heading at Q can be estimated by linear interpolation as follows:

$$\theta_r = (1 - \chi)\theta_k + \chi \theta_{k+1}, \tag{4.50}$$

where θ_k and θ_{k+1} are the heading angles at waypoint q_k and q_{k+1}, respectively. The heading error and lateral error for the discrete waypoint path are hence,

$$\Delta\psi = \psi - \theta_r, \tag{4.51}$$

$$e = \left|\overrightarrow{PQ}\right| = \frac{\left|\overrightarrow{q_k P} \times \overrightarrow{q_k q_{k+1}}\right|}{\left|\overrightarrow{q_k q_{k+1}}\right|}. \tag{4.52}$$

4.4.2 Tracking Error Computation for a Spline Path Representation

For the spline path representation, we similarly have to find the point Q on the path such that its tangent $\vec{t}$ is perpendicular to $\overrightarrow{PQ}$. After locating the nearest polynomial segment, the point Q can be located by finding the polynomial coefficient λ that minimizes the vehicle distance from the path,

$$\lambda = \arg\min_{\lambda \in [0,1]} \left|\overrightarrow{PQ}(\lambda)\right| = \arg\min_{\lambda \in [0,1]} \left\{ \left[x(\lambda) - x_P\right]^2 + \left[y(\lambda) - y_P\right]^2 \right\}, \tag{4.53}$$

or such that the dot product between $\overrightarrow{PQ}$ and the tangent $\vec{t}$ is zero:

$$\overrightarrow{PQ} \cdot \vec{t} = [x(\lambda) - x_P]\dot{x}(\lambda) + [y(\lambda) - y_P]\dot{y}(\lambda) = 0. \tag{4.54}$$

The value of λ that satisfies the above condition within the range $[0, 1]$ can be found by using the gradient descent or the Newton method. Once the point Q is located, the heading error and lateral error of the spline path become obvious:

$$\Delta\psi = \psi - \theta_r = \psi - a\tan 2(\dot{y}(\lambda), \dot{x}(\lambda)), \tag{4.55}$$

$$e = |\overrightarrow{PQ}| = \sqrt{[x(\lambda) - x_P]^2 + [y(\lambda) - y_P]^2}. \tag{4.56}$$

4.5 Chapter Summary and Concluding Remarks

In this chapter, we presented different path models ranging from use of discrete waypoints to parametric path descriptions. In the latter case, we compared approaches such as clothoids, Bezier curves, and polynomial spline descriptions and discussed their features and applications that are useful for autonomous driving. Specifically, the polynomial spline method was emphasized with our constrained least-square curve fitting technique to achieve smoothness and curvature continuity of the path. Tracking error calculation is necessary to generate steering feedback control action for accurate following of the modeled desired path. We presented two different calculation procedures for a discrete-waypoint path and for a spline path. The path generation and tracking error calculation methods presented in this chapter are necessary prerequisites for the material covered in Chapters 5–7 on path planning and path following control.

References

1 A. Savitzky and M. J. E. Golay, "Smoothing and differentiation of data by simplified least squares procedures," *Analytical Chemistry*, vol. 36, no. 8, pp. 1627–1639, 1964.

2 E. Bertolazzi and M. Frego, "G1 fitting with clothoids," *Journal of Mathematical Methods in the Applied Sciences*, vol. 38, no. 5, pp. 881–897, 2015.

3 A. Kelly and B. Nagy, "Reactive nonholonomic trajectory generation via parametric optimal control," *International Journal of Robotics Research*, vol. 22, no. 7–8, pp. 583–601, 2003.

4 H. Marzbani, R. N. Jazar, and M. Fard, "Better road design using clothoids," in *Sustainable Automotive Technologies 2014*, 2015, pp. 25–40: Springer.

5 B. Kufver, *Realigning Railways in Track Renewals-Linear Versus S-Shaped Superelevation Ramps*, Statens väg-och transportforskningsinstitut., VTI särtryck 318, 1999.

6 G. Orbay, M. E. Yümer, and L. B. Kara, "Sketch-based aesthetic product form exploration from existing images using piecewise clothoid curves," *Journal of Visual Languages and Computing*, vol. 23, no. 6, pp. 327–339, 2012.

7 J. Funke and J. Christian Gerdes, "Simple clothoid lane change trajectories for automated vehicles incorporating friction constraints," *Journal of Dynamic Systems, Measurement, and Control*, vol. 138, no. 2, pp. 021002-1–021002-9, 2016.

8 D. K. Wilde, "Computing clothoid segments for trajectory generation," in *2009 IEEE/RSJ International Conference on Intelligent Robots and Systems*, 2009, pp. 2440–2445: IEEE. doi:https://doi.org/10.1109/IROS.2009.5354700.

9 J. Funke, P. Theodosis, R. Hindiyeh, *et al.*, "Up to the limits: autonomous Audi TTS," *2012 IEEE Intelligent Vehicles Symposium*, Madrid, Spain, 2012, 541–547: IEEE, doi:https://doi.org/10.1109/IVS.2012.6232212.

10 C. Alia, T. Gilles, T. Reine, and C. Ali, "Local trajectory planning and tracking of autonomous vehicles, using clothoid tentacles method," *2015 IEEE Intelligent Vehicles Symposium (IV)*, Seoul, South Korea, 2015, pp. 674–679: IEEE, doi:https://doi.org/10.1109/IVS.2015.7225762.

11 Haui, "De Casteljau construction," 2005. Available: https://commons.wikimedia.org/wiki/File:De_Casteljau_construction_2.png#file.

12 K. Jolly, R. S. Kumar, R. J. R. Vijayakumar, and A. Systems, "A Bezier curve based path planning in a multi-agent robot soccer system without violating the acceleration limits," *Robotics and Autonomous Systems*, vol. 57, no. 1, pp. 23–33, 2009.

13 J. P. Rastelli, R. Lattarulo, and F. Nashashibi, "Dynamic trajectory generation using continuous-curvature algorithms for door to door assistance vehicles," in *2014 IEEE Intelligent Vehicles Symposium Proceedings*, 2014, pp. 510–515: IEEE.

14 D. González, J. Perez, R. Lattarulo, V. Milanés, and F. Nashashibi, "Continuous curvature planning with obstacle avoidance capabilities in urban scenarios," in *17th International IEEE Conference on Intelligent Transportation Systems (ITSC)*, 2014, pp. 1430–1435: IEEE.

15 J.-w. Choi, R. Curry, and G. Elkaim, "Path planning based on Bézier curve for autonomous ground vehicles," in *Advances in Electrical and Electronics Engineering-IAENG Special Edition of the World Congress on Engineering and Computer Science 2008*, 2008, pp. 158–166: IEEE.

16 Z. Liang, G. Zheng, and J. Li, "Automatic parking path optimization based on Bezier curve fitting," in *2012 IEEE International Conference on Automation and Logistics*, 2012, pp. 583–587: IEEE.

17 X. Qian, I. Navarro, A. de La Fortelle, and F. Moutarde, "Motion planning for urban autonomous driving using Bézier curves and MPC," in *2016 IEEE 19th International Conference on Intelligent Transportation Systems (ITSC)*, 2016, pp. 826–833: IEEE.

18 M. T. Emirler, "Advanced Control Systems for Ground Vehicles," PhD, İstanbul Technical University, 2014.

5

Collision Free Path Planning

This chapter starts with a detailed review of the literature on collision-free path planning methods available in the literature. The methods discussed for collision avoidance include the elastic band method, the quintic spline method with minimum curvature variation, and the model-based trajectory planning method. After a collision threat is determined by detecting and tracking obstacles, a path-planning module is used to plan a feasible, collision-free path that avoids possible collisions with these obstacles. The first method that is proposed deforms the path like an elastic band, using virtual forces exerted by the surrounding obstacles. The efficacy of this elastic band approach is demonstrated in both hardware-in-the-loop (HIL) simulations and experiments that handle both static and moving obstacles, pedestrians in the example that is presented. In the second method, the spline path is optimized, aiming at minimum curvature variation of the path for smoothness and passenger comfort, with computational efficiency being achieved by using lookup tables. In combination with a geometric method to generate collision-free target points in the grid map, the quintic spline method is seen to be capable of planning a smooth and safe path in realistic simulations. This chapter also covers the model-based trajectory planning method which adds another dimension of the longitudinal motion into planning by generating a set of candidate trajectories to choose from. Its performance is demonstrated on a low friction road, combined with the road friction coefficient estimation module from Chapter 3, and evaluated in different complex driving scenarios with multiple objects. The chapter ends with a discussion and comparison of the different methods introduced, followed by conclusions.

5.1 Introduction

Path planning is a challenging task for autonomous driving in a dynamically changing environment. The planned path should be collision-free with surrounding obstacles in the environment, while also smooth enough for the benefits of smooth path following and passenger comfort. A poorly designed path would not only make the subsequent path following difficult but may also cause passenger discomfort with unnecessarily large lateral acceleration and undesired yaw oscillation. Due to the complexity of the dynamically changing driving environment, the planned path generally requires the capability to adjust itself in real time to avoid sudden obstacles detected in its way.

The most important objective of safety-critical applications in autonomous and connected vehicles is to avoid crashes or collisions. This is usually done by modifying the autonomous vehicle path which requires collision free-path planning. One of the simpler approaches that can be used for path planning is the use of motion primitives. Although it is a very efficient method

Autonomous Road Vehicle Path Planning and Tracking Control, First Edition.
Levent Güvenç, Bilin Aksun-Güvenç, Sheng Zhu, and Şükrü Yaren Gelbal.

computationally [1], the use of a motion primitive by itself is not suitable for real vehicles because of the limitations it brings in terms of motion trajectories. It is mostly used in trajectory planning for mobile robots and unmanned air vehicles and also preferred for cooperative trajectory planning [2, 3] on these platforms, mainly because of its very efficient computation. Another approach that is widely used in robotics is rapidly exploring random trees (RRT). Unfortunately, when used by itself, randomness brings disadvantages such as large computation times and longer, nonoptimal and nonsmooth trajectories. Even though the RRT algorithm has been improved significantly through the years and specialized methods aimed for vehicles such as smooth RRT have been developed, computation times [4] and optimality of the obtained path are still existing issues. Model predictive control (MPC) is another, relatively more complex approach that can be used for both path tracking control and path planning. For the aspect of collision-free path determination, MPC is mostly used for overtaking and lane change type maneuvers in the literature. Reference [5], for example, proposes an active collision avoidance system that uses the MPC approach for predicting the collision and controlling the vehicle. Reference [6], on the other hand, uses the MPC method in combination with potential fields to determine an optimal overtaking trajectory. Definition of the many constraints and solving the cost minimization problem in MPC is relatively time-consuming.

The elastic band approach was first proposed in reference [7] as a mobile robotics-based method to create a deformable path that is subjected to artificial forces for the robots to move around the obstacles. This method carries similarities with the potential field method [8], and the main idea is changing the path dynamically while in motion where the forces are designed to have much smoother distribution. Since the path modification is constrained by a known, feasible, predefined path; the resulting motion is also feasible and predictable. The path modification is, thus, computationally fast and can be executed independently from the path following method. With developments and improvements of the method, the application cases have increased widely, and the method has become more suitable for collision-free path planning of automated vehicles [9]. It was used in the literature for different application cases such as enhancing vehicle following behavior [10], collision avoidance between vehicles in different traffic conditions [11], and recently for collision avoidance with vulnerable road users (VRUs) [12].

An alternative way is to use optimization methods to generate a smooth path in the two-dimensional plane. However, its high computation burden limits its direct application to online path planning. The optimization problem to minimize the curvature variation of a quintic polynomial path is first raised. It adopts a time-efficient online table-lookup approach to reduce the optimization computation cost. Given discrete target points, this approach is capable of forming a quintic spline path with second-order geometric (G^2) continuity with the aid of a look-up table. G^2 continuity, also known as curvature continuity, means that the second derivative and, hence, the curvature has the same value at each point of the path. This means that the second derivative will not change whether we approach the point form one direction or the opposite direction. Similarly, a G^0 continuous path is continuous and has no discrete jumps and a G^1 continuous path has the same first derivative at each point regardless of the direction of approach.

The optimization-based table is generated offline and beforehand in a so-called "reference space" so that its required dimensions can be reduced. In the reference space where initial and target endpoints are fixed, lookup tables of optimal path parameters that minimize the curvature variation are constructed offline using an optimization method at different sets of orientation angles and curvature values. In online implementation, two endpoints in each segment are transformed into the reference space where linear interpolations of the lookup tables are completed. An inverse transformation is used afterwards to obtain the optimal path with G^2 continuity in real space. In this way, lookup tables of up to four dimensions are sufficient within a small storage space.

This is then combined with a geometry-based target points generation method to form a collision-free path to avoid obstacles.

Since both the elastic-band approach and the polynomial path planning have no inherent longitudinal motion planning and lack the consideration of vehicle dynamics, state and control constraints, if the area of application requires such features, the trajectory optimization approach can be used as an alternative that is capable of weighing the relative importance of each objective within imposed constraints of kinematics and obstacles, hence, providing a multiobjective solution [13–15].

The trajectory planning problem generally aims to find a feasible connection from the initial state to reach a target state. This type of problem has been thoroughly investigated in the field of robotic motion planning, hence, naturally, several methods from which were applied to autonomous vehicles. These methods often rely on discretized state structures, including sampling-based and graph search-based planners [16]. Of the sampling-based techniques, RRT and its variant RRT* were mostly seen in applications to full-sized vehicles [17–19]. The graph search-based techniques including state lattices [20, 21] and A* algorithms [22, 23] were also commonly applied. These methods are quite favored for combinatorially difficult problems as encountered in unstructured environments such as parking lots. However, they are usually incapable of handling multiple alternative goal states which may be necessary for an evasive maneuver. Their computational complexity prevents them from being applied in fast-changing dynamic environments. Their sparse control discretization during the search would also lead to passenger discomfort for on-road driving from medium to high speed.

These features make the above strategies unsuitable for on-road driving scenarios. On-road trajectory planning should place road safety as its top priority with the capability of planning a collision-free trajectory, while also considering passenger comfort and the use of road structure information. Some existing work in the literature focused on the vehicle's reactive capability to avoid other road users (ORU) or obstacles. In [24, 25], the artificial potential field was constructed by superposition of the obstacles and the roadmap with its descent direction for forming the collision-safe path. In [26, 27], MPC was applied with properly formulated obstacle constraints and, hence, integrated trajectory planning and tracking. In [13], a local trajectory optimization was conducted continuously with obstacle shapes constrained as polygons to achieve dynamic feasibility and comfort. Other approaches including the elastic band method [28] and the support vector machine [29] were less commonly used and also aimed at similar reactive capability. These reactive approaches work fine for light traffic by themselves but are difficult to integrate with an upper behavior planner to handle more complex driving scenarios such as merging into another lane for example. Some other on-road planning techniques, on the other hand, focused on a particular scenario such as overtaking [30, 31] or merging [32] and are too scenario-specific and lack ease of adaptivity to other driving conditions.

We therefore embark on the phased pipeline proposed in [33] which took advantage of the road structure by deliberately sampling multiple alternative final states over the local horizon which maintains the reactive capability for collision avoidance by itself. Moreover, it could work under an integrated framework where its upper behavior layer could assign desired longitudinal and lateral goal states. Designation of multiple final states is not new. Polynomials [33, 34] and Bezier curves [35, 36] were used to generate the trajectory candidates to reach these final states due to their simplicity and smoothness. However, these trajectories might not be dynamically feasible as only curve fitting techniques were applied without considering the vehicle dynamics and constraints. In [15, 37], a trajectory optimization problem bounded by state terminals at both ends was formed with the state equations describing the vehicle movement. The curvature control was parameterized as a cubic polynomial to facilitate the optimization efficiency.

During the state trajectory planning, we also took advantage of the parameterized control optimization but with a more versatile curvature representation. The trajectory optimization problem with spline-parameterized curvature control was formulated to reach the goal state with the vehicle model and its dynamic constraints considered. This in contrast to the existing curve fitting techniques that guarantee the dynamic feasibility of the planned trajectory. With a generation of multiple trajectory candidates along the Frenet frame, the vehicle is reactive to ORU or obstacles encountered. Moreover, the decoupling of longitudinal and lateral motion planning enables the trajectory planner to respond to the desired goal state requests from its upper behavior layer to deal with complex driving scenarios. The potential of the proposed trajectory planning framework was demonstrated under different dynamic driving scenarios such as lane-changing or merging with surrounding vehicles with its computational efficiency demonstrated in real-time simulations. The main benefits of this alternative method are a computationally efficient optimization-based trajectory planning with guaranteed dynamically feasibility, and its seamless integration with the behavior layer demonstrated by complex dynamic driving scenarios such as real-time lane changing and merging. Based on the discussion in this section, some of the advantages and disadvantages of the different path planning algorithms used for path planning and collision avoidance that were discussed above are summarized in Table 5.1.

The rest of this chapter after this literature review explores the different approaches ranging from the elastic band method, the quintic spline with minimum curvature variation method, to the model-based trajectory planning method. After the introduction of the theoretical background and feature of each method, results from simulations, HIL tests, and vehicle experiments under different scenarios are given to demonstrate their effectiveness.

Table 5.1 Advantages and disadvantages of some path planning methods.

Method	Advantages	Disadvantages	Computation time
Motion primitives [5]	Very simple and very fast	Route is neither optimal nor flexible. Behavior can be unnatural for a car	Low
Rapidly exploring random trees [8]	Easy to implement. Suitable for holonomic kinematics	Multiple random searches are relatively time-consuming. Can cause unnatural behavior for a car due to randomness	Medium–high
Model predictive control [6]	On-line path generation and modification are combined	With constraints and weights, calculation of optimization is relatively slow	Medium–high
Elastic band [10]	Modified path is well defined and structured. Transition from or to the non-modified path is natural. Calculation is relatively fast	Relies on a different algorithm for generation of pre-determined path	Low–medium
Optimization [38]	Capable of including constraints such as other road users and the road environment into consideration	Convergence time can be high for complex problems. Global minimization not guaranteed for nonconvex problem	Medium–high

5.2   Elastic Band Method

Path planning is one of the most important aspects of the collision avoidance behavior in autonomous driving. Obtaining a safe and comfortable maneuver path is often a hard task that requires one to take many different arguments into account. The elastic band method is a very flexible and modular method that can be used for localized path modification in collision avoidance, which also provides a reasonable amount of comfort and safety. In this section, along with the path structure, the elastic band method is explained with force calculation and equilibrium equations and demonstrated with several examples of path modifications involving static obstacles.

5.2.1   Path Structure

The predefined path is the main route that the vehicle is expected to follow unless there are any interventions. This path is described by discrete waypoints that are equidistant. These waypoints can be generated based on GPS points or using a predefined path geometry and a series of smooth polynomial curve fits, as described in Chapter 4. Waypoints are localized around the start point of the path and represented with X and Y locations in meters. Figure 5.1 shows the plotted waypoints for a path generated from GPS points collected from a straight manual drive. Each waypoint is represented as a circle. The path starts with the first waypoint at $x = 0$ (m) and $y = 0$ (m) and ends with the 113th waypoint at $x = -111.61$ (m) and $y = 8.91$ (m).

5.2.2   Calculation of Forces

The elastic band is a modifiable band that has different types of potential fields and resulting forces acting on it. The types of forces to be used depend on the application area or the scenario. In this study, for collision avoidance, we focused on two main forces, internal forces caused by nodes themselves and external forces caused by obstacles like other vehicles and vulnerable road users such as pedestrians, bicyclists, and scooters. The nodes exert the internal on each other as a pull force that holds the band together and provides its elasticity. External forces are applied by the obstacles on the nodes as push forces and deform the path for making it collision-free. Each waypoint on the path is assigned to be a node of the elastic band in the vicinity of the obstacles. The internal and external forces acting on the nodes are illustrated in Figure 5.2 where a pedestrian is viewed as the obstacle with potential collision to the original path.

In Figure 5.2, while the vehicle is following the predefined path consisting of nodes $i = 1, 2, 3, 4, \ldots$ marked with their corresponding position vectors $\mathbf{u_i}$, an obstacle, shown as a pedestrian in Figure 5.2, is encountered and external forces are applied on the nodes of the path to deform it into a collision-free path. Some of the nodes such as node 4 with position vector $\mathbf{u_4}$ and node 5 with

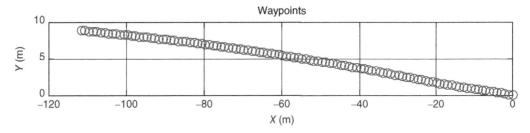

Figure 5.1   Generated waypoints for a manually driven straight path.

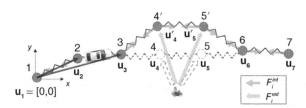

Figure 5.2 Internal and external forces acting on elastic band.

position vector $\mathbf{u}_5$ in Figure 5.2 are affected by this force more than others and their displacements are larger. The path is held together by the internal forces while the nodes are displaced locally around the obstacle. At the end of all the iterations for force and displacement calculations, the nodes arrive at their new positions marked as node i' with new position vector $\mathbf{u}'_i$. This iteration process is repeated at each steering controller computation time step while running in real-time. Therefore, even though the obstacle may be moving in a cross path, the deflected vehicle path is smoothly and continuously modified for each obstacle location at each time t which results in a smooth transition between paths and allows the vehicle to behave naturally while following the continuously evolving modified path.

For any given elastic band, the position vector of each node i is defined as

$$\mathbf{u_i} = \begin{bmatrix} x_i \\ y_i \end{bmatrix}, \tag{5.1}$$

and the Euclidean norm between adjacent nodes i and $i-1$ is defined as

$$\|\mathbf{u_i} - \mathbf{u_{i-1}}\| = \sqrt{(y_i - y_{i-1})^2 + (x_i - x_{i-1})^2}. \tag{5.2}$$

For each node position vector $\mathbf{u_i}$, the distance $d_{i,p}$ from the previous node $i-1$ and the distance $d_{i,n}$ to the next node $i+1$ are defined as

$$d_{i,p} = \|\mathbf{u_i} - \mathbf{u_{i-1}}\|, \tag{5.3}$$

$$d_{i,n} = \|\mathbf{u_i} - \mathbf{u_{i+1}}\|. \tag{5.4}$$

The internal potential energy $V^{int}(x_i, y_i)$ of each node i is due to the two springs it interconnects. If this potential energy is viewed as a function of position coordinates x and y, i.e. $V^{int}(x,y)$, it is defined only at the node locations i with coordinates (x_i,y_i): $i = 1,2,...,n$ and is zero elsewhere. $V^{int}(x_i,y_i)$ is due to the deflection energy of the two springs connected to node i and is given by

$$V_i^{int}(x_i, y_i) = \frac{1}{2} \left[k_s(d_{i,p} - l_{0,p})^2 + k_s(d_{i,n} - l_{0,n})^2 \right], \tag{5.5}$$

or equivalently using Eqs. (5.3) and (5.4) as

$$V^{int}(x_i, y_i) = \frac{1}{2} \left[k_s(\|\mathbf{u_i} - \mathbf{u_{i-1}}\| - l_{0,p})^2 + k_s(\|\mathbf{u_i} - \mathbf{u_{i+1}}\| - l_{0,n})^2 \right], \tag{5.6}$$

where k_s is spring coefficient, $l_{0,p}$ is initial (unstretched) length from the previous node, and $l_{0,n}$ is initial (unstretched) length to the next node. Both $l_{0,p}$ and $l_{0,n}$ are constant values. In the artificial potential field method, a potential field $V(x,y)$ is defined for all points (x,y) within the area of operation. In contrast, the internal potential field $V^{int}(x_i,y_i)$ of the elastic band method is only calculated at a small, finite number of nodes. The artificial potential field $V(x,y)$ is calculated once and is independent of the vehicle path. It only changes if the obstacles in the operating area move. On the other hand, the elastic band nearby nodes, where the calculations are made and their internal potential field $V^{int}(x_i,y_i)$ are recalculated at each time interval since they change as the vehicle moves forward along its path.

Using Eq. (5.6), the internal force $\mathbf{F_i^{int}}(x_i, y_i)$ acting on each node i can be calculated by using the negative gradient of the potential function as

$$\mathbf{F^{int}}(x_i, y_i) = -\nabla V^{int}(x_i, y_i) = -\left(\frac{\partial}{\partial x} i + \frac{\partial}{\partial y} j \right) V^{int}(x, y) \Bigg|_{(x,y)=(x_i, y_i)}, \tag{5.7}$$

$$\mathbf{F^{int}}(x_i, y_i) = -k_s(\|\mathbf{u_i} - \mathbf{u_{i-1}}\| - l_{0,p}) \underbrace{\frac{\mathbf{u_i} - \mathbf{u_{i-1}}}{\|\mathbf{u_i} - \mathbf{u_{i-1}}\|}}_{\substack{\text{unit vector} \\ \text{along } (i-1)\text{th spring}}} + k_s(\|\mathbf{u_{i+1}} - \mathbf{u_i}\| - l_{0,n}) \underbrace{\frac{\mathbf{u_{i+1}} - \mathbf{u_i}}{\|\mathbf{u_{i+1}} - \mathbf{u_i}\|}}_{\substack{\text{unit vector} \\ \text{along } (i+1)\text{th spring}}}. \tag{5.8}$$

For the external forces, obstacle locations can be used to define potentials generated by the obstacles on each node. At any given time t, if the obstacle position vector is defined as $\mathbf{u_{obs}}(t)$, the distance between the obstacle and the ith node of the elastic band, $d_{i,obs}$ is,

$$d_{i,obs}(t) = \|\mathbf{u_i} - \mathbf{u_{obs}}(t)\|. \tag{5.9}$$

The potential created by an obstacle $V^{obs}(x_i, y_i)$ is a repulsive potential field with logarithmic relation to the distance and is given by

$$V^{obs}(x_i, y_i) = k_{obs} \ln(\|\mathbf{u_i} - \mathbf{u_{obs}}(t)\|), \tag{5.10}$$

where k_{obs} is the repulsive potential coefficient for the obstacle. The external force caused by the obstacle $\mathbf{F^{obs}}(x_i, y_i)$ on the band is calculated as the negative gradient of the potential field in Eq. (5.12) as

$$\mathbf{F^{obs}}(x_i, y_i) = -\nabla V^{obs}(x_i, y_i) = \frac{k_{obs}(\mathbf{u_i} - \mathbf{u_{obs}}(t))}{\|\mathbf{u_i} - \mathbf{u_{obs}}(t)\|^2}. \tag{5.11}$$

In addition to the traditionally calculated force, the coefficient k_{obs} in Eq. (5.11) has been made dynamic depending on the minimum distance between the obstacle and the nodes on the path. This was done so that the repulsive force can be completely zeroed out after some distance, as we are not interested in obstacles who are more than a certain distance away from the path. This definition also allows relatively smooth transition between the zeroed out coefficient and the maximum value as the obstacle gets closer. The coefficient is defined as

$$k_{obs} = \begin{cases} k_{obs,max} - \frac{d_{min}(t)}{5}, & k_{obs,max} > \frac{d_{min}(t)}{5} \\ 0, & k_{obs,max} < \frac{d_{min}(t)}{5} \end{cases}, \tag{5.12}$$

where $k_{obs,max}$ is the maximum value for the coefficient and $d_{min}(t)$ is the distance of the closest node to the obstacle at time t. Finally, the external force acting on the ith node, $\mathbf{F^{ext}}(x_i, y_i)$, can be defined as the summation of the forces generated by all the obstacles as

$$\mathbf{F^{ext}}(x_i, y_i) = \sum_{j=1}^{M} \mathbf{F^{obs_j}}(x_i, y_i), \tag{5.13}$$

where $\mathbf{F^{obs_j}}(x_i, y_i)$ is the repulsive force generated by jth obstacle on the ith node (x_i, y_i) and M is the number of obstacles.

5.2.3 Reaching Equilibrium Point

With the forces defined in Section 5.2.2, the total force acting on each node can be written as the sum of external and internal forces as

$$\mathbf{F}^{\text{sum}}(x_i, y_i) = \mathbf{F}^{\text{int}}(x_i, y_i) + \mathbf{F}^{\text{ext}}(x_i, y_i) = 0, \tag{5.14}$$

where the sum is assigned to be zero corresponding to force equilibrium being reached. In order to solve this equilibrium, a numerical iteration method was implemented. In each iteration, small displacements are applied on each node as a fraction of the resulting force as

$$\mathbf{u_i'} = \mathbf{u_i} + k_f \mathbf{F}^{\text{sum}}(x_i, y_i), \tag{5.15}$$

where k_f is a small number. Forces are recalculated after the displacements are applied and new displacements are applied according to the newly calculated forces. This iteration continues until forces approach equilibrium. In other words, until

$$\mathbf{F}^{\text{sum}}(x_i, y_i) \cong 0. \tag{5.16}$$

This approach allows limiting the number of iterations, which, when tuned, can obtain results very close to the equilibrium point much faster than an analytical solution. Some examples of path modifications for a straight path with random obstacle (pedestrian) locations are shown in Figure 5.3 for the described implementation of the elastic band method. Different configurations

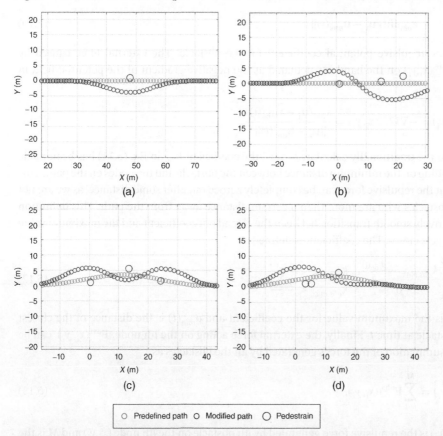

Figure 5.3 Some path modification examples for different path and pedestrian placement configurations: (a) single obstacle (pedestrian) – straight path, (b) multiple obstacles (pedestrians) – straight path, (c, d) multiple obstacles (pedestrians) – curved path.

Table 5.2 Algorithm for modification of vehicle path nodes.

Algorithm: Node Location Modification

Initialization

initialize parameters k_s, k_{obs}, k_f

for all node pairs:

  calculate l_0 values (eqs. 5.1–5.4)

end

for all obstacles:

  calculate closest distance to path (eq. 5.9)

  if obstacle is on the path:

    move the obstacle by ε

  end

end

Main Loop

while equilibrium is not reached $(F_{sum} > \varepsilon)$

  calculate internal forces (eqs. 5.5–5.8)

  calculate external forces (eqs. 5.10–5.13)

  calculate the amount of displacement caused by sum of forces (eqs. 5.14, 5.15)

  move the nodes

end

of paths (straight, curved), different obstacle counts, and location configurations can be seen along with successful path modifications in Figure 5.3. The almost straight line of circles represents the predefined vehicle path and the line of curved circles represents the collision avoidance trajectory generated with the elastic band method (Table 5.2).

5.2.4 Selected Scenarios

After the implementation of the elastic band method as presented above, three different scenarios which are obtained from high crash risk scenarios involving pedestrians as determined by National Highway Traffic Safety Association (NHTSA) [39] are studied next, in order to evaluate the collision avoidance capability of the elastic band approach. These scenarios are selected to be realistic evaluations and demonstrations that support the purpose of pedestrian safety. The scenarios are illustrated on Figure 5.4 with moving pedestrians and vehicles.

The first scenario, scenario A, is a parked vehicle scenario where there are two vehicles that are parked on the side of the road. The pedestrian suddenly moves toward the road between these parked vehicles, while the ego vehicle is driving on the road toward the pedestrian's direction. In the manual driving case, this scenario would start as a Non-Line-of-Sight (NLOS) scenario because of the two parked vehicles obstructing the view of the pedestrian. After the pedestrian moves out of the space between those vehicles, it will be too late for the driver to react and a crash will most likely happen. This scenario can be generalized to other environments where vehicles are parked close to each other like driveways, parking lots with a lot of parked cars, and many pedestrians walking around. The second scenario, scenario B, is a crash scenario where the pedestrian is jogging at the side of the road in the same direction as the ego vehicle. The pedestrian might not notice the vehicle and make lateral movements, or the driver might not be able to notice the pedestrian under reduced visibility conditions [40]. The third scenario, scenario C, is a right turn scenario, where a

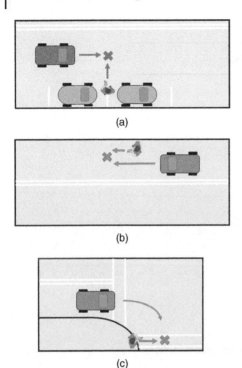

(a)

(b)

(c)

Figure 5.4 Illustrations of selected NHTSA scenarios A (sudden appearance between parked cars), B (pedestrian walking alongside the road), and C (pedestrian in way of right turning car).

pedestrian is attempting to cross the road at the same time as the ego vehicle is executing a sudden right turn maneuver. Especially if the driver is hurrying, he/she might not notice the pedestrian and attempt to make a right turn which would result in a possible crash.

Realistic representations for each scenario were created for the HIL simulations and road experiments. While creating the representations of these scenarios, availability of the infrastructure in the testing grounds and possible variations of the elements used in testing the algorithm, such as number of pedestrians, motion of pedestrians, and road shape need to be taken into consideration to evaluate the algorithm as thoroughly as possible. It is also important to emphasize that these scenarios assume the situation of emergency, where the vehicle is not allowed to or cannot stop in time to prevent these crashes.

5.2.5 Results

For each scenario presented in Section 5.2.4, a HIL simulation and experiment were conducted to analyze the performance. A picture of the automated vehicle at the testing location used can be seen in Figure 5.5. Pedestrians in the simulations and real world are virtually created pedestrians that communicate their location through V2X connection. Experiments are done within the testing grounds with no traffic and at a speed of 20 km/h. Each test is explained starting with the description and illustration of the representation for the scenario, followed by the vehicle paths plotted on top of the satellite image, performance results of HIL simulation, and the road experiment, followed by a discussion of results.

Test scenario A was created in a parking lot area. Two pedestrians were assumed to be leaving their cars, walking between other parked cars and standing close to the driving path/predefined path of the ego vehicle. An illustration of the test scenario setup is shown in Figure 5.6. Pedestrians

Figure 5.5 Picture of the experimental vehicle at testing grounds.

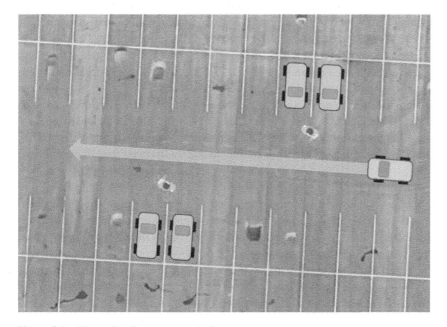

Figure 5.6 Illustration for test scenario A.

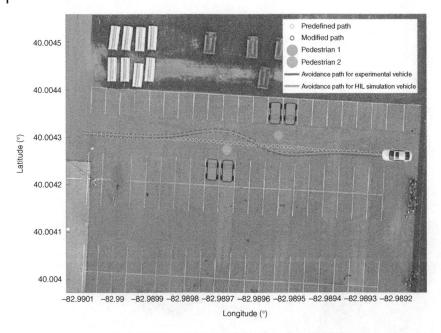

Figure 5.7   HIL and experimental vehicle paths for scenario A testing.

can be seen after they have parked their vehicles, appearing close to the road between parked cars and standing dangerously close to the ego vehicle's predefined path.

This experiment involves two static pedestrians, and the vehicle is expected to maneuver around them safely. Path tracking results for the HIL and real-world experiments can be seen in Figure 5.7. The vehicle path for both HIL simulation and the real-world experiment is plotted aside each other and can be compared with the modified path. As seen in Figure 5.7, the vehicle is successfully able to modify its path and safely maneuver around the pedestrians. Moreover, the HIL experiment and real-world results are also seen to closely match each other.

Lateral deviations from the overall avoidance path are also analyzed for both HIL simulation and the experiment and shown in Figure 5.8. Since the vehicles start slightly out of the path, there is an offset error initially. Rest of the time, both errors can be seen to be mostly staying under 0.2 m with root mean square (RMS) values of 0.1871 m for the experiment and 0.1104 m for the HIL simulation results.

Test scenario B involves a pedestrian jogging or walking at the side of the road in the same direction as the ego vehicle. The ego vehicle is approaching from behind the pedestrian, and since the pedestrian is very close to the path, it is expected to avoid and maneuver around the pedestrian. The scenario is illustrated in Figure 5.9. The pedestrian can be seen walking dangerously close to the nearby vehicle's predefined path.

A straight pathway is used for this test. There is a single pedestrian moving at around 1 m/s speed at the same direction with the vehicle. Multiple pedestrian locations with respect to time after the pedestrian starts moving are also marked, with the corresponding modified path at that certain time instant. These pedestrian locations and corresponding modified paths can be seen along with the avoidance path of the vehicle for HIL and experiment, in Figure 5.10.

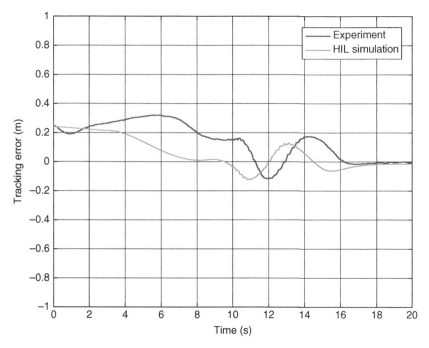

Figure 5.8 Tracking errors for scenario A testing.

Figure 5.9 Illustration for test scenario B.

As seen in Figure 5.10, sometime after the pedestrian starts moving along the road, the vehicle starts to stray away from the predefined path in order to avoid the pedestrian. While doing this, both the vehicle and the pedestrian are in motion and a new modified path is calculated very frequently in real-time, according to the pedestrian location at that certain time. Only a couple of these paths are shown in Figure 5.10 for better readability, but it can be seen from Figure 5.10 that the vehicle makes smooth transitions between these dynamically changing paths, while successfully avoiding the pedestrian in both HIL simulation and real-world experiment. The final path that the vehicle follows is formed by switching to the next calculated path at each computation instance and is shown by the straight lines in Figure 5.10 which are the HIL simulation and real-world experiment results. They are seen to be closely matching with each other. Tracking performance for the HIL simulation and experiment are plotted together in Figure 5.11.

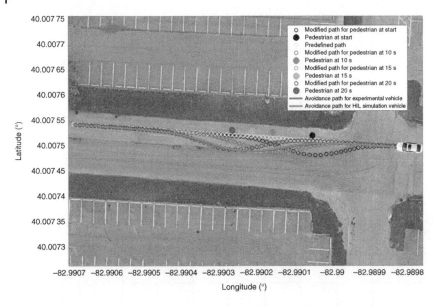

Figure 5.10 HIL and experimental vehicle paths for scenario B testing.

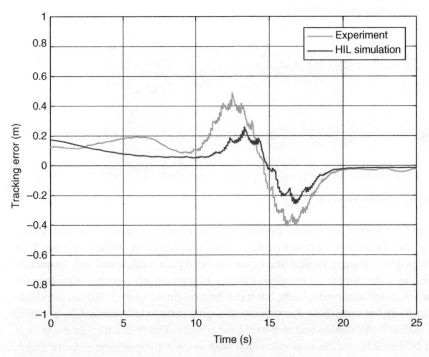

Figure 5.11 Tracking errors for scenario B testing.

Figure 5.12 Illustration for test scenario C.

Similar to the previous scenario, there is a slight offset from the path at the start. But for the rest of the path, tracking error mostly stays under 0.2 m. RMS error values are 0.1893 m for the experiment and 0.1019 m for the HIL simulation.

Test scenario C involves a curved predefined path where the ego vehicle is planning to turn right. A pedestrian suddenly moves very close to the path of the vehicle with the intent of crossing. This results in the ego vehicle maneuvering around the pedestrian to avoid the crash. Illustration of the test scenario can be seen in Figure 5.12. The ego vehicle is expected to turn right while there is a pedestrian standing very close to the predefined vehicle path and intending to cross the road.

A static virtual pedestrian was put around the crosswalk near the vehicle path, for testing. The vehicle was successfully able to avoid the pedestrian in both HIL simulation and the experiment. Both results are plotted in Figure 5.13.

It can be seen that the predefined path is modified into a smooth avoidance maneuver path and the vehicle is able to avoid the pedestrian safely. The maneuver in this scenario also slightly represents a double lane change maneuver around the curved predefined path which actually can be realized more closely if the elastic band coefficients are tuned accordingly. Tracking errors for both results are shown in Figure 5.14.

Starting positions are much closer to the path compared to previous test scenarios. The rest of the path is also tracked pretty well with around 0.12 m error at most. RMS error value for the experiment is 0.0469 m and for the HIL simulation is 0.0449 m. All of the RMS values for the discussed scenarios are summarized in Table 5.3.

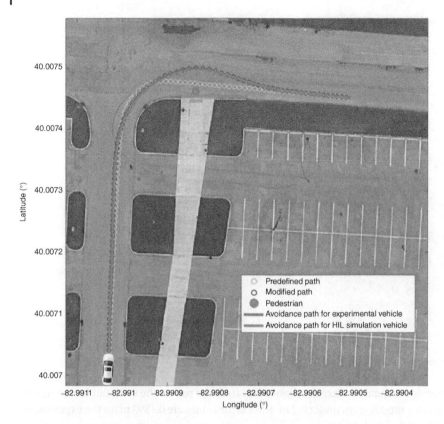

Figure 5.13 HIL and experimental vehicle paths for scenario C testing.

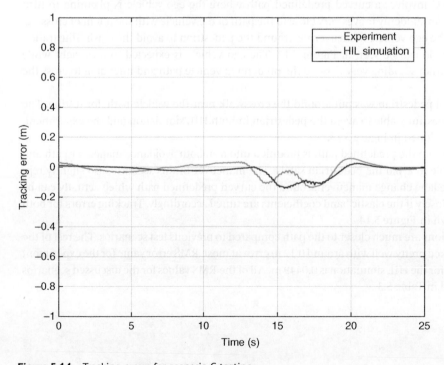

Figure 5.14 Tracking errors for scenario C testing.

Table 5.3 RMS values for tracking error.

Test scenarios	HIL (m)	Experiment (m)
Scenario A	0.1104	0.1871
Scenario B	0.1019	0.1893
Scenario C	0.0449	0.0469

5.3 Path Planning with Minimum Curvature Variation

5.3.1 Optimization Based on G^2-Quintic Splines Path Description

In this section, the path in the two-dimensional plane is described by a polynomial spline with each segment expressed as a quintic polynomial $\mathbf{p}(\lambda)$, $\lambda \in [0, 1]$ given by

$$\mathbf{p}(\lambda) = \begin{bmatrix} x(\lambda) \\ y(\lambda) \end{bmatrix} := \begin{bmatrix} x_0 + x_1\lambda + x_2\lambda^2 + x_3\lambda^3 + x_4\lambda^4 + x_5\lambda^5 \\ y_0 + y_1\lambda + y_2\lambda^2 + y_3\lambda^3 + y_4\lambda^4 + y_5\lambda^5 \end{bmatrix}, \tag{5.17}$$

where x_i, y_i, $(i = 0, 1, \ldots, 5)$ are polynomial coefficients to be determined. To determine these coefficients for each segment, the position $\mathbf{p}$, orientation $\hat{\mathbf{e}}$, and curvature κ information of the two endpoints are required (Figure 5.15):

$$\mathbf{p}(0) = \mathbf{p}_A, \tag{5.18a}$$

$$\mathbf{p}(1) = \mathbf{p}_B, \tag{5.18b}$$

$$\hat{\mathbf{e}}(0) = \begin{bmatrix} \cos\theta_A \\ \sin\theta_A \end{bmatrix}, \tag{5.18c}$$

$$\hat{\mathbf{e}}(1) = \begin{bmatrix} \cos\theta_B \\ \sin\theta_B \end{bmatrix}, \tag{5.18d}$$

$$\kappa(0) = \kappa_A, \tag{5.18e}$$

$$\kappa(1) = \kappa_B. \tag{5.18f}$$

It has been shown in [42] that if the polynomial coefficients could be expressed with respect to $\boldsymbol{\eta} := [\eta_1\ \eta_2\ \eta_3\ \eta_4]^T \in \Omega := (0, \infty) \times (0, \infty) \times (-\infty, \infty) \times (-\infty, \infty)$ as follows, then the interpolation constraints are always fulfilled for any $\boldsymbol{\eta} \in \Omega$:

$$x_0 = x_A, \tag{5.19a}$$

$$x_1 = \eta_1 \cos\theta_A, \tag{5.19b}$$

Figure 5.15 Interpolation conditions of the endpoints. Source: Zhu et al. [41] (© [2018] IEEE).

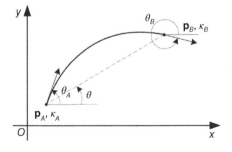

$$x_2 = \frac{1}{2}\left(\eta_3 \cos \theta_A - \eta_1^2 \kappa_A \sin \theta_A\right), \tag{5.19c}$$

$$x_3 = 10(x_B - x_A) - \left(6\eta_1 + \frac{3}{2}\eta_3\right)\cos \theta_A - \left(4\eta_2 - \frac{1}{2}\eta_4\right)\cos \theta_B$$
$$+ \frac{3}{2}\eta_1^2 \kappa_A \sin \theta_A - \frac{1}{2}\eta_2^2 \kappa_B \sin \theta_B, \tag{5.19d}$$

$$x_4 = -15(x_B - x_A) + \left(8\eta_1 + \frac{3}{2}\eta_3\right)\cos \theta_A$$
$$+ (7\eta_2 - \eta_4)\cos \theta_B - \frac{3}{2}\eta_1^2 \kappa_A \sin \theta_A + \eta_2^2 \kappa_B \sin \theta_B, \tag{5.19e}$$

$$x_5 = 6(x_B - x_A) - \left(3\eta_1 + \frac{1}{2}\eta_3\right)\cos \theta_A - \left(3\eta_2 - \frac{1}{2}\eta_4\right)\cos \theta_B$$
$$+ \frac{1}{2}\eta_1^2 \kappa_A \sin \theta_A - \frac{1}{2}\eta_2^2 \kappa_B \sin \theta_B, \tag{5.19f}$$

$$y_0 = y_A, \tag{5.20a}$$

$$y_1 = \eta_1 \sin \theta_A, \tag{5.20b}$$

$$y_2 = \frac{1}{2}\left(\eta_3 \sin \theta_A + \eta_1^2 \kappa_A \cos \theta_A\right), \tag{5.20c}$$

$$y_3 = 10(y_B - y_A) - \left(6\eta_1 + \frac{3}{2}\eta_3\right)\sin \theta_A - \left(4\eta_2 - \frac{1}{2}\eta_4\right)\sin \theta_B$$
$$- \frac{3}{2}\eta_1^2 \kappa_A \cos \theta_A + \frac{1}{2}\eta_2^2 \kappa_B \cos \theta_B, \tag{5.20d}$$

$$y_4 = -15(y_B - y_A) + \left(8\eta_1 + \frac{3}{2}\eta_3\right)\sin \theta_A + (7\eta_2 - \eta_4)\sin \theta_B$$
$$+ \frac{3}{2}\eta_1^2 \kappa_A \cos \theta_A - \eta_2^2 \kappa_B \cos \theta_B, \tag{5.20e}$$

$$y_5 = 6(y_B - y_A) - \left(3\eta_1 + \frac{1}{2}\eta_3\right)\sin \theta_A - \left(3\eta_2 - \frac{1}{2}\eta_4\right)\sin \theta_B$$
$$- \frac{1}{2}\eta_1^2 \kappa_A \cos \theta_A + \frac{1}{2}\eta_2^2 \kappa_B \cos \theta_B. \tag{5.20f}$$

In this way, the quintic polynomial curve in (5.17) is reduced to four design freedoms with the parameter $\eta \in \Omega$ given in the interpolation constraints. The optimization problem is formulated as minimizing the maximum curvature variation:

$$\min_{\eta \in \Omega}\left(\max_{\lambda \in [0,1]}\left|\frac{d\kappa}{ds}(\lambda)\right| + w \cdot s(1)\right), \tag{5.21}$$

subject to

$$\|\dot{\mathbf{p}}(\lambda)\| > 0, \quad \forall \lambda \in [0,1], \tag{5.22a}$$

$$\eta_1 - \eta_2 = 0, \tag{5.22b}$$

$$\eta_3 + \eta_4 = 0, \tag{5.22c}$$

where $\|\dot{\mathbf{p}}(\lambda)\|$ is the 2-norm of the curve's gradient, and s in (5.21) is the curvilinear length measured along the path $\mathbf{p}(\lambda)$:

$$s(\lambda) = \int_0^\lambda \|\dot{\mathbf{p}}(\zeta)\| \, d\zeta. \tag{5.23}$$

The latter item in the objective function (5.21) adds penalty to the planned path length $s(1)$ weighted by w. A curve is said to be regular if it never slows to a stop or backtracks on itself, i.e. its derivative is never zero or negative. It is for this reason that the constraint (5.22a) is added to guarantee the regularity of the curve.

It is revealed in [42] of the relationship between the parameters η and the rate of change of curvilinear length s (or $ds/d\lambda$) with respect to λ:

$$\eta_1 = \frac{ds}{d\lambda}(0) = \|\dot{\mathbf{p}}(0)\|, \tag{5.24a}$$

$$\eta_2 = \frac{ds}{d\lambda}(1) = \|\dot{\mathbf{p}}(1)\|, \tag{5.24b}$$

$$\eta_3 = \frac{d^2 s}{d\lambda^2}(0) = \frac{d}{d\lambda}\|\dot{\mathbf{p}}(0)\|, \tag{5.24c}$$

$$\eta_4 = \frac{d^2 s}{d\lambda^2}(1) = \frac{d}{d\lambda}\|\dot{\mathbf{p}}(1)\|. \tag{5.24d}$$

Constraints ((5.22b) and (5.22c)) are added to expect symmetric distribution of $ds/d\lambda$ in $\lambda \in [0, 1]$ intuitively and simplify the solution to the above optimization problem. $\mathbf{p}$ is the position vector, $\dot{\mathbf{p}}$ is velocity, $ds/d\lambda$ and $\|\mathbf{p}\|$ are speed, and $d^2 s/d\lambda^2$ and $d\|\mathbf{p}\|/d\lambda$ are magnitudes of acceleration.

The minimization of curvature variation is set as the objective in order to reduce the steering variation since the steering angle δ_f is directly related to the traveled curvature κ for a single-track vehicle model given by

$$\delta_f = \left(L + k_{us} V_x^2\right) \kappa, \tag{5.25}$$

for steady state cornering, where L is the vehicle wheelbase, k_{us} is the understeer gradient coefficient determined by vehicle mass distribution and tire cornering stiffness coefficients, and V_x is the vehicle longitudinal speed. Therefore, a planned path with less curvature variation is desirable for smoother steering and less lateral acceleration perturbation.

The min–max optimization problem of Eq. (5.14) can be transformed to a semi-infinite programming form as seen in [42]. The solution to this problem can be pursued by converting semi-infinite constraints to a set of finite constraints with peak values and then using sequential quadratic programming (SQP). However, the solution to this semi-infinite programming problem comes with heavy calculation load and is hence not time efficient. It may also end up in local optima, and a solution is sometimes not guaranteed if the initial condition is poorly selected. These negative factors together limit its direct application in online path planning even though the path generated is ideal with minimum curvature variation.

5.3.2 Reduction of Computation Cost Using Lookup Tables

The collision-avoidance scenario is extremely time-sensitive and demands a quick path replanning especially at high speed. However, the optimization approach is not time-efficient since the optimization of the objective function takes place at each motion state [16]. The complexity of the optimization problem greatly increases if the optimal objective and constraints imposed become complicated. For this reason, online implementation of optimization is still restricted with a trade-off between explicit representation and computational efficiency, and usually requires a high-performance in-vehicle PC for calculation [43, 44]. Meanwhile, the global optimum is difficult to find for a nonconvex optimization problem. The optimization solution generally reduces to a local optimum in this case and may require another path selector to decide whether the planned path is acceptable or not [44, 45].

Instead of online optimization, path planning is usually divided into two phases as the smooth path generation phase followed by the collision-freeness evaluation phase [46]. The optimization cost in the former phase can be reduced by precomputing lookup tables of path parameters offline. For example, [15, 47] parameterized control inputs generated a model-based trajectory from initial states to target ones via the 5D precomputed tables. The parameterization of these control inputs, however, is sometimes not inclusive [47].

Due to the difficulty of direct online implementation of the optimization method, the table-lookup approach is raised with the idea to obtain suboptimal path parameters from the lookup tables constructed beforehand with extensive offline programming. However, the different combination of coordinates, orientation, and curvature values of the two end points (Figure 5.15) requires a look-up table of up to six dimensions with almost infinite choice of values.

To reduce the complexity of dimensionality and save storage space, a reference space is formed where coordinates of endpoints A' and B' are fixed, for example at (0,0) m and (10,0) m, respectively. Lookup tables of path parameters were constructed with respect to four-dimensional inputs of orientation angle and curvature values of the endpoints. The orientations $\theta_{A'}$ and $\theta_{B'}$ are defined in the range of $[-\pi/4, \pi/4]$ rad, and the curvature $\kappa_{A'}$ and $\kappa_{B'}$ in the range of $[-0.1, 0.1]$, each with an evenly divided grid of 20 numbers. The semi-infinite optimization problem (5.21) was solved offline at each combination of the defined orientations and curvature values. Figure 5.16 shows the results of path trajectories, curvature, and curvature variation value along the path at five different combinations of $\theta_{A'}$, $\theta_{B'}$, $\kappa_{A'}$, and $\kappa_{B'}$. Larger ranges of these lookup parameter values are unnecessary since the original endpoints problem can be divided with more medium points to fit inside the range of $\theta_{A'}$, $\theta_{B'}$, $\kappa_{A'}$, and $\kappa_{B'}$ after transformation into the reference space.

From these optimization solutions, the path parameter η' instead of the polynomial coefficients, x_i' and y_i' ($i = 0, \ldots, 5$) were stored into a total of four four-dimensional (4D) lookup tables to reduce table size.

Figure 5.17 shows the table-lookup path planning procedure. A transformation of the original endpoints into the reference space is required first. The constraints, i.e. orientation angles and curvature values, of the transformed endpoints A' and B' are then referred to the 4D lookup tables to obtain parameter values η' through linear interpolation. This path found in the reference space is then reflected back to the original coordinate system through the de-transformation process.

Transformation
To reflect the endpoints A and B into their counterparts A' and B' fixed in the reference space, coordinate scaling and rotation are required. The orientation angles and curvature values of the endpoints after transformation are

$$\theta_{A'} = \theta_A - \theta, \tag{5.26a}$$

$$\theta_{B'} = \theta_B - \theta, \tag{5.26b}$$

$$\kappa_{A'} = \kappa_A \cdot \frac{\|AB\|}{\|A'B'\|}, \tag{5.26c}$$

$$\kappa_{B'} = \kappa_B \cdot \frac{\|AB\|}{\|A'B'\|}, \tag{5.26d}$$

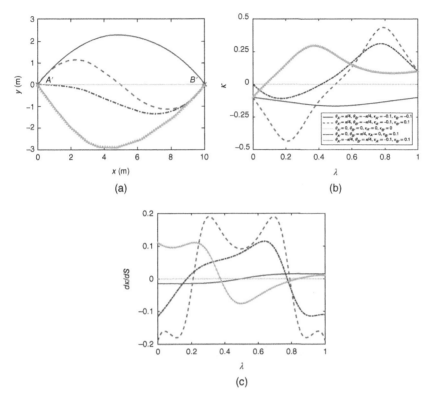

Figure 5.16 (a) Optimal polynomial curve, (b) curvature $\kappa(\lambda)$, and (c) curvature variation $d\kappa/ds(\lambda)$ for some orientation angles and curvature of endpoints A′ and B′ in the reference space (legend shown in (b)). Source: Zhu et al. [41] (© [2018] IEEE).

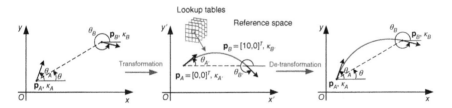

Figure 5.17 Table-lookup path planning procedure. Source: Zhu et al. [41] (© [2018] IEEE).

where θ is the angle of line AB with respect to the abscissa axis (Figure 5.17), and the Euclidean distance $\|AB\|$ and $\|A'B'\|$ are used to scale the curvature values into the reference space.

Table Lookup

The $\theta_{A'}$, $\theta_{B'}$, $\kappa_{A'}$, and $\kappa_{B'}$ values after transformation are fed to the 4D lookup tables to get the path parameter values of η' through linear interpolation. The linear interpolation includes merely simple algebraic calculation and is not detailed here. Polynomial coefficients, x_i' and y_i' ($i = 0$, 1, …, 5), can then be determined from parameter η' using ((5.19a)–(5.19f)) and ((5.20a)–(5.20f)), respectively.

De-transformation

A de-transformation is necessary to reflect the polynomial coefficients, x_i' and y_i' ($i = 0, 1, ..., 5$), in the reference space back to the original coordinate systems. The polynomial curve in the real space has the relation below with its counterpart in the reference space:

$$\mathbf{p}(\lambda) = \begin{bmatrix} x(\lambda) \\ y(\lambda) \end{bmatrix} = \begin{bmatrix} x_A \\ y_A \end{bmatrix} + \frac{\|AB\|}{\|A'B'\|} \cdot \mathbf{R}_\theta \begin{bmatrix} x_0' + x_1'\lambda + x_2'\lambda^2 + x_3'\lambda^3 + x_4'\lambda^4 + x_5'\lambda^5 \\ y_0' + y_1'\lambda + y_2'\lambda^2 + y_3'\lambda^3 + y_4'\lambda^4 + y_5'\lambda^5 \end{bmatrix}, \tag{5.27}$$

where $\mathbf{R}_\theta$ is the rotation matrix

$$\mathbf{R}_\theta = \begin{bmatrix} \cos\theta & -\sin\theta \\ \sin\theta & \cos\theta \end{bmatrix}. \tag{5.28}$$

Equation (5.27) can be reorganized as the following matrix form with respect to the vector of $[1\ \lambda \cdots \lambda^5]^T$ as

$$\mathbf{p}(\lambda) = \begin{bmatrix} x_A + \frac{\|AB\|}{\|A'B'\|}\left((x_0' + x_1'\lambda + x_2'\lambda^2 + x_3'\lambda^3 + x_4'\lambda^4 + x_5'\lambda^5)\cos\theta - \right. \\ y_A + \frac{\|AB\|}{\|A'B'\|}\left((x_0' + x_1'\lambda + x_2'\lambda^2 + x_3'\lambda^3 + x_4'\lambda^4 + x_5'\lambda^5)\sin\theta + \right. \\ \left(y_0' + y_1'\lambda + y_2'\lambda^2 + y_3'\lambda^3 + y_4'\lambda^4 + y_5'\lambda^5\right)\sin\theta) \\ \left(y_0' + y_1'\lambda + y_2'\lambda^2 + y_3'\lambda^3 + y_4'\lambda^4 + y_5'\lambda^5\right)\cos\theta) \end{bmatrix}, \tag{5.29}$$

$$= \begin{bmatrix} x_A + \frac{\|AB\|}{\|A'B'\|}(x_0'\cos\theta - y_0'\sin\theta) & \frac{\|AB\|}{\|A'B'\|}(x_1'\cos\theta - y_1'\sin\theta) \\ y_A + \frac{\|AB\|}{\|A'B'\|}(x_0'\sin\theta + y_0'\cos\theta) & \frac{\|AB\|}{\|A'B'\|}(x_1'\sin\theta + y_1'\cos\theta) \end{bmatrix}$$

$$\begin{bmatrix} \frac{\|AB\|}{\|A'B'\|}(x_1'\cos\theta - y_1'\sin\theta) & \cdots & \frac{\|AB\|}{\|A'B'\|}(x_5'\cos\theta - y_5'\sin\theta) \\ \frac{\|AB\|}{\|A'B'\|}(x_1'\sin\theta + y_1'\cos\theta) & \cdots & \frac{\|AB\|}{\|A'B'\|}(x_5'\sin\theta + y_5'\cos\theta) \end{bmatrix} \begin{bmatrix} 1 \\ \lambda \\ \vdots \\ \lambda^5 \end{bmatrix}. \tag{5.30}$$

The original path expression with respect to parameters, x_i and y_i ($i = 0, 1, ..., 5$), is in similar form:

$$\mathbf{p}(\lambda) = [\mathbf{M}]_{2\times 6}\begin{bmatrix} 1 & \lambda & \cdots & \lambda^5 \end{bmatrix}^T, \tag{5.31}$$

where $\mathbf{M}$ is the 2×6 matrix which contains the polynomial coefficients x_i and y_i ($i = 0, 1, ..., 5$):

$$\mathbf{M} = \begin{bmatrix} x_0 & x_1 & x_2 & x_3 & x_4 & x_5 \\ y_0 & y_1 & y_2 & y_3 & y_4 & y_5 \end{bmatrix}. \tag{5.32}$$

By relating Eqs. (5.29) and (5.32), the values of the path parameters, x_i and y_i ($i = 0, 1, ..., 5$), for the original two endpoints can be found.

Figure 5.18 shows that the suboptimal solution of the table-lookup path planning approximates very close to the optimization result with endpoints given at $\mathbf{p}_A = [11, 99]^T$, $\mathbf{p}_B = [-10, -10]^T$, $\theta_A = -86.5°$, $\theta_B = -117.2°$, $\kappa_A = 0$, and $\kappa_B = 0$ as an example. Similar results are obtained at other endpoints conditions, validating the applicability of the table-lookup approach in path planning. Besides, it takes much shorter time to implement, i.e. 7.3 μs on average in Simulink Desktop Real-Time™ environment on a computer with 2.60 GHz Intel i7-5600U processor and 12 GB RAM, compared to semi-infinite programming (Table 5.4).

5.3.3 Geometry-Based Collision-Free Target Points Generation

The proposed table-lookup path planning requires the information of discrete target points. These preview target points need to be generated dynamically to provide the vehicle the freedom to avoid obstacles in real time. There are several ways of generating these target points, such as elastic

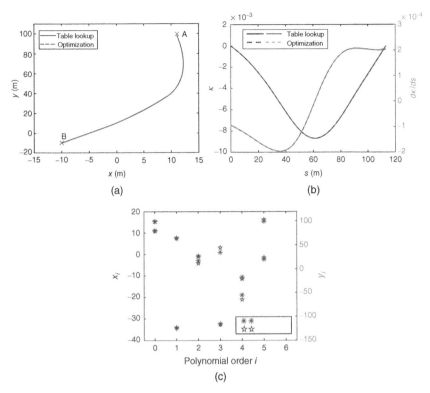

Figure 5.18 Comparison of table-lookup approach and optimization solution: (a) path trajectory; (b) curvature κ and its variation $d\kappa/ds$; (c) polynomial coefficients x_i and y_i. Source: Zhu et al. [41] (© [2018] IEEE).

Table 5.4 Time cost comparison.

	Semi-infinite programming	Table-lookup approach
Time cost	178.8 ms	7.3 μs

band [12, 48] and potential field methods [45, 49]. These two methods maintain the vehicle at a relative safe distance from the obstacle based on the defined elastic force or potential field. However, the absolute distance of the path from the obstacles are generally uncertain in the planning phase when these methods are used. Depending on the tuned algorithm parameters, the generated target points could be too conservative or too close to the obstacles, neither of which are desirable for a collision-avoidance behavior. For this reason, a geometry-based method is proposed with ascertained absolute distance from obstacles and have the benefit of low computation cost.

Collision Prediction

A two-dimensional local occupancy grid map needs to be formed first with each grid cell size of $0.5\,\mathrm{m} \times 0.5\,\mathrm{m}$. In the real-world case, this grid map can be formed using sensors such as camera, radar, and LIDAR. Information from these sensors can be used to update the map continuously while the vehicle is on the move, with the use of different packages available in robot operating system (ROS), for example. Some of the packages that can be used are the simultaneous localization

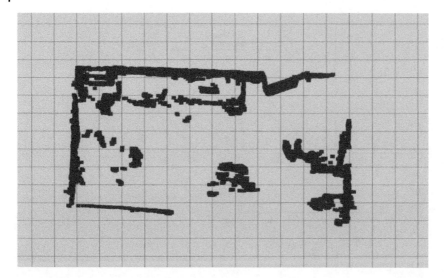

Figure 5.19 ROS visualization of LIDAR data with surrounding environment. Source: Zhu et al. [41] (© [2018] IEEE).

and mapping (SLAM) algorithm [50, 51] packages such as the Gmapping and Hector SLAM packages. A visualization of the LIDAR data that can be used for grid mapping is shown in Figure 5.19 as an example, with data recorded with a Velodyne 16-channel LIDAR. This sensor is used in real-world implementation and experiments in other work of the authors.

The planned path assumes the vehicle as a point mass. To account for the size of the vehicle itself to avoid collisions, the grid areas of the objects in the grip map are enlarged to provide some conservativeness. The controller calculates a preview path trajectory with 100 m look-ahead distance and judges if potential collision is expected. The collision prediction algorithm first evaluates the grid cells where the future path goes through and then checks if any of these pass-through grid cells coincide with any cells that the obstacles occupy. If any coincidence is observed as seen in Figure 5.20, potential collision would be expected and a new path is in need to avoid the obstacle.

Collision-Free Target Points Generation

If potential collision is expected through the collision prediction module, then new collision-free target points need to be generated. Assuming P and Q are the corners where the future path enters and exits the obstacle (Figure 5.21), the start and end target points A and B can be found in the original path with a travel distance s before P and after Q, respectively:

$$s = \max \{V_x \Delta t, 5\,\text{m}\}, \tag{5.33}$$

where Δt is a time constant determined by the user and 5 m is the minimum allowable reserved distance. Notice that in actual application, the decision-making module also needs to be designed and decide whether to stop or pass the obstacle considering other traffic. The longitudinal speed profile should be designed accordingly. This section mainly focuses on online path planning so decision-making and speed profile design are not covered here.

The intermediate target points are found as follows. The idea is to shift line PQ parallelly with step size 0.5 m until a collision-free parallel line l is found (Figure 5.21) [52] and directly chose a single target point E closest to the final parallel line l to form a two-segment evasive path together with points A and B. However, if the obstacle is not convex as in the case of Figure 5.21, the planned path

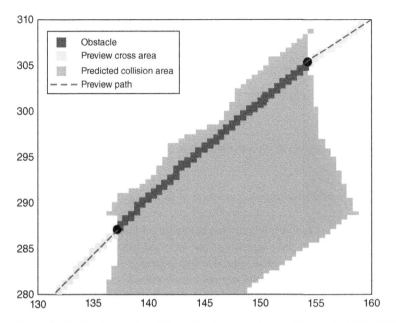

Figure 5.20 Preview path collides with obstacles. Source: Zhu et al. [41] (© [2018] IEEE).

Figure 5.21 Illustration of collision-free target points generation. Source: Zhu et al. [41] (© [2018] IEEE).

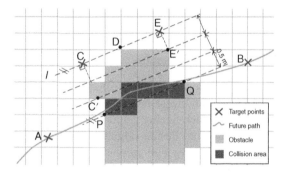

A–D–B might still collide with the obstacle and requires another or more path planning procedures afterward. These extra computations can be avoided by designing a three-segment evasive path instead. The idea is to find the widest potential collision points C′ and E′ along the shift direction and reflects these two points to the parallel line *l* as C and E. The evasive path A–C–E–B is then ascertained to avoid the obstacle with one computation.

Besides the locations of the collision-free target points, orientation angles and curvature information at these points are still needed before linear interpolation is carried out using the lookup tables generated in Section 5.3.2 to form the path. The orientation and curvature values for points A and B are known from the original planned path. For points C and E, orientation angles are set to the angle of line *l* with respect to the abscissa axis and curvature values are set to zero in order to form a straight path segment CE.

Online Path Planning and Update

Given the target points A, B, C, and E generated with the geometry-based method, each segment could be formed with the two neighboring endpoints through the table-lookup procedure: transformation, lookup, and de-transformation (Figure 5.17). The three segments together form a spline

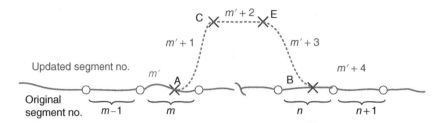

Figure 5.22 Original and updated path segments along with their orders. Source: Zhu et al. [41] (© [2018] IEEE).

with second-order geometric (G^2) continuity, because curvature at the endpoints and in between are both continuous.

To update the path information, the neighboring segments before target point A and after point B are also re-planned because the original path segments (the mth and nth segment shown in Figure 5.22) were broken apart by A and B. Therefore, a total of five segments are constructed with the table-lookup online path planning method. The five segments are inserted between the original $(m-1)$th and $(n+1)$th segments to form the new path with updated segment order (Figure 5.22).

The updated path information is then used to form a new preview path trajectory. The process of collision prediction and re-planning is then continued based on the new preview path. Therefore, dynamically moving objects and multiple obstacles can be taken into account with the proposed table-lookup online path planning approach.

5.3.4 Simulation Results

Simulations were carried out in Simulink to validate the table-lookup online path planning technique along with the geometry-based target points generation method in the scenario of collision avoidance. A high-fidelity vehicle model in CarSim with parameters customized to match our experimental Ford Fusion vehicle was used during the simulation. The scenario was created at a typical road curve with an obstacle in the original planned path for the vehicle. The obstacle has an irregular shape in order to evaluate the capability of the online path planning to decide an appropriate surpass distance.

Figure 5.23 shows the result of the aforementioned online path planning technique with the original and updated paths, actual vehicle path, and the obstacle highlighted in an occupancy grid map. The vehicle followed its original planned path and online path planning was inactive until the vehicle reached the star position (Figure 5.23) where the preview path was found to be in collision with the detected obstacle. Four new targets labeled in cross (Figure 5.23) were generated based on the obstacle shape with the geometry-based method along with their orientation and curvature information. These target points were interpolated into the lookup tables generated before to form a collision-free path with minimum curvature variation. A total of five new segments in solid line (Figure 5.23) were constructed with the table-lookup approach and in replacement of the original path along with an updated segment order.

Figure 5.24 shows a zoom-in of Figure 5.23 where the area across the original preview path and the predicted collision area with the obstacle are highlighted. It can be seen that the two generated intermediate target points have accounted for the widest potential collision zone in the path shift direction with the proposed geometry-based method. The vehicle path also shows good following performance with steering control at the speed of 15 m/s.

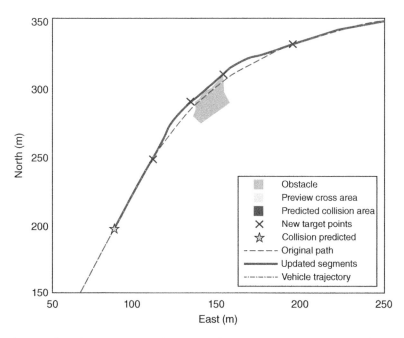

Figure 5.23 Online path planning result with an obstacle at a road curve. Source: Zhu et al. [41] (© [2018] IEEE).

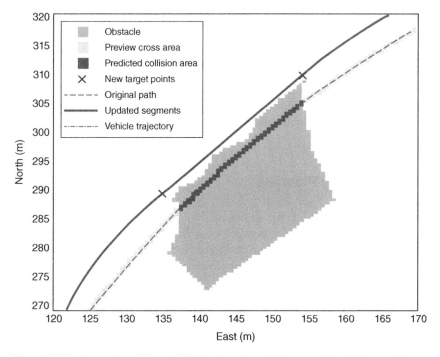

Figure 5.24 A zoom-in of Figure 5.23 showing the widest potential collision points were accounted for in an irregular-shaped obstacle with the geometry-based method. Source: Zhu et al. [41] (© [2018] IEEE).

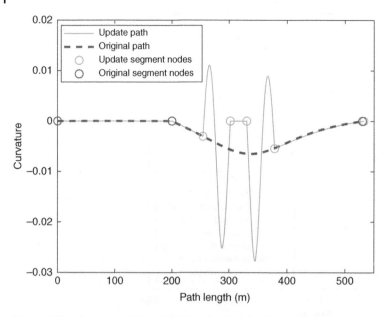

Figure 5.25 Curvature of the original and updated path and their corresponding segment nodes. Source: Zhu et al. [41] (© [2018] IEEE).

The updated path shows a smooth transition from the original path to the updated segments to evade the obstacle (Figure 5.23). Curvature change is continuous along the updated path (Figure 5.25), with a minimization of curvature variation ensured by the priori generated optimization-approach-based lookup tables. The smooth curvature change is desired as it directly affects steering as seen in Eq. (5.25). The segment nodes before and after path updating are also shown in Figure 5.25 to illustrate the path-updating process in Figure 5.22.

Figure 5.26 shows the steering and path following performance for the collision avoidance scenario. The steering along with yaw rate and lateral acceleration is smooth, thanks to the G^2-continuity of the updated path, which improves passenger comfort. Besides, the lateral error and heading angle error are within acceptable range, validating the effectiveness of the designed path following controller.

5.4 Model-Based Trajectory Planning

There are certain limitations revealed during the two-dimensional geometric path planning as presented in Section 5.3, including the lack of consideration of vehicle dynamics, state, and control constraints. It also does not plan the longitudinal movement for the vehicle. Due to these concerns, we have further explored the state trajectory planning in this section to lift these limitations.

5.4.1 Problem Formulation

A general description of the trajectory generation problem is to find a feasible control $\mathbf{u}$ such that the state trajectory $\mathbf{x}$ satisfies the terminal constraints [37], as expressed in

$$\mathbf{x}(\zeta_f) = \mathbf{x}_f, \tag{5.34}$$

$$\text{s.t. } \dot{\mathbf{x}}(\zeta) = f(\mathbf{x}, \mathbf{u}, \zeta), \mathbf{x}(\zeta_0) = \mathbf{x}_0, \mathbf{x} \in \mathbf{X}, \text{and } \mathbf{u} \in \mathbf{U}. \tag{5.35}$$

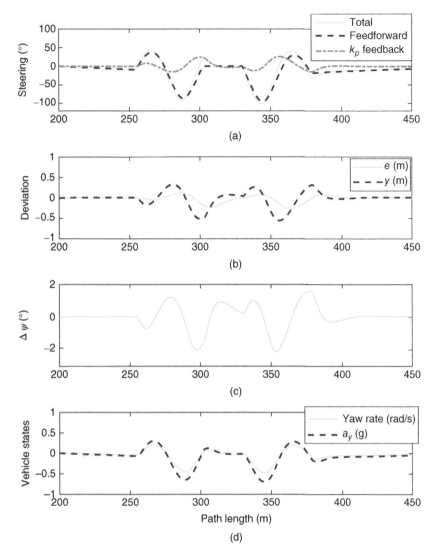

Figure 5.26 (a) Steering control, (b) lateral and look-ahead error, (c) heading angle error, (d) vehicle yaw rate and lateral acceleration during the collision avoidance at 15 m/s. Source: Zhu et al. [41] (© [2018] IEEE).

ζ denotes the independent variable of which typical choices can be time t or arc-length s which are interchangeable. The set of control actions satisfying the terminal constraint requirement might be infinite. Hence, a utility functional can be added besides Eqs. (5.34) and (5.35) to select a desired control action among all candidates:

$$\min_{\mathbf{u} \in U} J(\mathbf{x}, \mathbf{u}, \zeta). \tag{5.36}$$

The model $\dot{\mathbf{x}}(\zeta) = f(\mathbf{x}, \mathbf{u}, \zeta)$ in (5.35) describes the state trajectory, i.e. the vehicle motion, given the input $\mathbf{u}$. Selection of this model is a trade-off between description accuracy and computational efficiency. In this section, a simple dynamic model similar to other works [15, 53] was used to represent the vehicle motion due to its low computation complexity while still having enough accuracy at small tire slip angles:

$$\dot{x}(t) = v(t)\cos\,\theta(t), \tag{5.37}$$

$$\dot{y}(t) = v(t)\sin\,\theta(t), \tag{5.38}$$

$$\dot{\theta}(t) = v(t)\kappa(t), \tag{5.39}$$

where the states $\mathbf{x} = [x,\ y,\ \theta]^T$ are vehicle coordinates and heading, the control $\mathbf{u} = [\kappa,\ v]^T$ are velocity and curvature, respectively.

The model in Eqs. (5.37)–(5.39) retains relatively good accuracy given that the tire slip angles are small, which could be ensured by adding control constraints in the planning stage. More complex vehicle models like bicycle or four-wheel dynamic models could reflect the vehicle motion more accurately but at the cost of increased computation complexity which limits their direct online implementation. Additionally, if the wheel dynamics is also included to expand the model's applicability to highly nonlinear region of the tires, an infinitesimal discretization time step during the numerical integration is required to accommodate the fast-changing wheel dynamics which makes the planning impractical at best [54].

Another benefit of using the model in Eqs. (5.37)–(5.39) besides its simplicity is that the differential equations can be easily reorganized with respect to the arc-length s as follows:

$$dx/ds = \cos\,\theta(s), \tag{5.40}$$

$$dy/ds = \sin\,\theta(s), \tag{5.41}$$

$$d\theta/ds = \kappa(s), \tag{5.42}$$

which decouples the velocity v and relates the vehicle motion to the control of curvature κ solely. This typical approach simplifies the trajectory planning process by planning control v and κ separately [15, 55].

5.4.2 Parameterized Vehicle Control

The control over the horizon $[\zeta_0, \zeta_f]$ being optimal in the sense of the constrained optimization problem (5.34)–(5.36) cannot be solved in its analytical form in general when applying the variational approach or Pontryagin's minimum principle. Numerical approaches like the gradient projection method to solve this problem with state and control constraints are feasible but could be time-consuming if trying to optimize $\mathbf{u}(k)$ at every discretized time step [56]. Therefore, the control over the horizon could be parameterized with respect to fewer free parameters, thus reducing the number of parameters to be optimized.

In this section, the controls for the trajectory planning, i.e. the curvature and velocity, are parameterized over the arc-length and time, and with respect to parameters $\boldsymbol{\rho}$ and $\boldsymbol{\gamma}$, respectively:

$$\mathbf{u} \triangleq [\kappa(\boldsymbol{\rho}, s), v(\boldsymbol{\gamma}, t)]^T. \tag{5.43}$$

While the parameterization for controls may confine the searching scope to only a subspace of all feasible motions, an appropriate selection of the parameterization can describe nearly all possible controls for the application. In this section, the curvature is represented as a cubic spline consisting of several polynomial segments. The polynomial coefficients in each segment could be determined by the spline nodes in between with C^1 continuity constraints. Therefore, the shape of the spline could be defined by just a few spline nodes, as shown in Figure 5.27. The spline parameterization

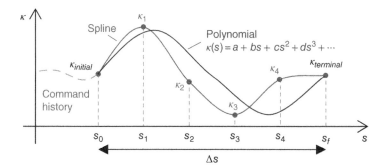

Figure 5.27 Parameterization of the curvature control with a spline versus a polynomial. Source: Zhu and Aksun-Guvenc [57]/with permission from Springer Nature.

covers more searching scope for the curvature control as compared to the single polynomial representation used in [37]. Figure 5.27 illustrates the comparison of the two parameterized controls in a typical lane-change maneuver. It is apparent that the spline representation provides more flexibility while allowing the curvature back to steady once lane change is finished. The single polynomial shape however is highly dependent on the terminal length s_f which makes it less consistent if s_f is subject to change. If the terminal length is too large, the polynomial representation could lead to an unpleasantly slow maneuver.

Compared to the curvature, the velocity is far less changeable in the planning horizon, which makes simple parameterization such as a trapezoidal profile [53] or a single polynomial [58] applicable with enough smoothness. Due to the decoupling of curvature and velocity control in the motion model (5.40)–(5.42), we could separate the planning for both and leave the longitudinal motion design for later discussion.

5.4.3 Constrained Optimization on Curvature Control

As mentioned previously, the curvature control is parameterized as a cubic spline with evenly distributed spline nodes and the arc-length and given by $\boldsymbol{\rho} = [\kappa_1, \kappa_2, \ldots, \kappa_p, s_f]^T$. The number of spline nodes p is chosen as 4 as it is found to be enough to describe all possibilities in a relative short planning horizon. Due to the parameterization, the control constraints for the curvature such as saturation and rate limits now have to be applied implicitly to the spline before obtaining the final feasible curvature $\kappa(\boldsymbol{\rho}, s)$:

$$|\kappa| \leq \min \left\{ \frac{\delta_{f,max}}{L + k_{us}v^2}, \frac{a_{y,max}}{v^2} \right\}, \tag{5.44}$$

$$\left| \frac{d\kappa}{ds} \right| \leq \frac{\dot{\delta}_{f,max}}{L + k_{us}v^2}, \tag{5.45}$$

where $\delta_{f,max}$ and $\dot{\delta}_{f,max}$ are the saturation and rate limit of the front steering wheel angle, $a_{y,max}$ is the maximum allowed lateral acceleration capped by the road friction coefficient μ:

$$a_{y,max} = \sigma \mu g, \tag{5.46}$$

where the safety coefficient $\sigma = 0.67$ is chosen considering the maximum steerability by the front tires for a front-drive vehicle. The trajectory evaluation module covered later will evaluate whether the combined longitudinal and lateral acceleration values exceed limits. L is the wheelbase and k_{us}

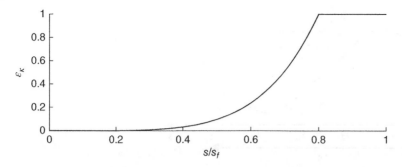

Figure 5.28 Distribution of error coefficient ε_k in favor of early curvature stabilization. Source: Zhu and Aksun-Guvenc [57]/with permission from Springer Nature.

is the understeer coefficient [59] as before. v in Eqs. (5.44) and (5.45) takes a constant value of the current velocity v_0 since the planning horizon is relatively short.

Since the curvature is parameterized with respect to parameters ρ, the original optimal control formulation (5.34)–(5.36) becomes a constrained optimization problem with respect to the free parameters ρ:

$$\min_{\rho} J(\rho), \tag{5.47}$$

where the terminal constraint is

$$\mathbf{c}(\rho) = \mathbf{x}(s_f, \rho) - \mathbf{x}_f = 0, \tag{5.48}$$

and the motion model is given in Eqs. (5.40)–(5.42) and the initial condition $\mathbf{x}(s_0) = \mathbf{x}_0, \mathbf{x} \in \mathbf{X}$, and curvature constraints in Eqs. (5.44) and (5.45) are satisfied.

The utility functional is a description of what we want to minimize over the path. Here we choose it as follows:

$$J(\rho) = \int_{s_0}^{s_f} \left\{ w_1[\varepsilon_k \cdot (\kappa(\rho, s) - \kappa_f)]^2 + w_2|a_y|^3 \right\} ds, \tag{5.49}$$

where the first item in the integrand favors the control that stabilizes itself toward the end curvature value κ_f as early as possible, with ε_k the error coefficient to distribute the penalty along the arc-length ratio s/s_f (Figure 5.28). Early here means achieving the final curvature value not exactly at the end point but at some normalized distance (like 20% in Figure 5.28) before the end point. The latter item adds the penalty cubically to increasing lateral acceleration. The utility functional is hence a trade-off between fast maneuver and lateral comfort decided by the weight values w_1, w_2.

The SQP approach is one of the most effective methods for nonlinearly constrained optimization problems like Eqs. (5.47) and (5.49). We begin by defining the Lagrangian as the sum of the cost function with the product of the Lagrange multiplier vector λ and the constraints.

$$\mathcal{L}(\rho, \lambda) = J(\rho) + \lambda^T \mathbf{c}(\rho) = J(\rho) + \lambda^T(\mathbf{x}(s_f, \rho) - \mathbf{x}_f). \tag{5.50}$$

The idea behind the SQP approach is to model (5.47) and (5.48) at each iteration ρ_k to a much simpler quadratic programming (QP) subproblem and then to use the minimizer of this subproblem for defining a new iterate ρ_{k+1}:

$$\min_{\mathbf{d}} J_k + \nabla J_k^T \mathbf{d} + \frac{1}{2}\mathbf{d}^T \nabla_{\rho\rho}^2 \mathcal{L}_k \mathbf{d}, \tag{5.51}$$

$$\text{s.t. } \nabla \mathbf{c}_k \mathbf{d} - \mathbf{c}_k = 0, \tag{5.52}$$

where $\mathbf{d}$ is the descent direction of $\boldsymbol{\rho}_k$. The minimizer $\mathbf{d}$ for this subproblem can be obtained to fulfill the following first-order necessary conditions:

$$\begin{bmatrix} \nabla_{\mathbf{pp}}^2 \mathcal{L}_k & \nabla \mathbf{c}_k^T \\ \nabla \mathbf{c}_k & 0 \end{bmatrix} \begin{bmatrix} \mathbf{d}^* \\ \boldsymbol{\mu}^* \end{bmatrix} = \begin{bmatrix} -\nabla J_k \\ \mathbf{c}_k \end{bmatrix}, \tag{5.53}$$

where $\boldsymbol{\mu}$ is the Lagrange multiplier introduced for the QP subproblem (5.51) and (5.52).

The gradient ∇J_k and $\nabla \mathbf{c}_k$ in the above equations cannot be derived analytically. Forward or central differences are used to obtain their numerical estimation values. As the order of the differentials increases, the finite difference estimate becomes less accurate. Therefore, the Hessian of the Lagrangian $\nabla_{\mathbf{pp}}^2 \mathcal{L}_k(\boldsymbol{\rho}_k, \lambda_k)$ in (5.51) is replaced by a quasi-Newton approximation B_k with damped Broyden–Fletcher–Goldfarb–Shanno (BFGS) updating [56]:

$$B_{k+1} = B_k - \frac{B_k s_k s_k^T B_k}{s_k^T B_k s_k} + \frac{r_k r_k^T}{s_k^T r_k}, \tag{5.54}$$

where

$$s_k = \boldsymbol{\rho}_{k+1} - \boldsymbol{\rho}_k, \tag{5.55}$$

$$y_k = \nabla_{\mathbf{p}} \mathcal{L}(\boldsymbol{\rho}_{k+1}, \lambda_{k+1}) - \nabla_{\mathbf{p}} \mathcal{L}(\boldsymbol{\rho}_k, \lambda_k), \tag{5.56}$$

$$\theta_k = \begin{cases} 1 & \text{if } s_k^T y_k \geq 0.2 s_k^T B_k s_k \\ \left(0.8 s_k^T B_k s_k\right) / \left(s_k^T B_k s_k - s_k^T y_k\right) & \text{if } s_k^T y_k < 0.2 s_k^T B_k s_k \end{cases}, \tag{5.57}$$

$$r_k = \theta_k y_k + (1 - \theta_k) B_k s_k. \tag{5.58}$$

Once the minimizer of the QP subproblem is obtained through (5.53), the new iterate $\boldsymbol{\rho}_k$ are updated as

$$\boldsymbol{\rho}_{k+1} = \boldsymbol{\rho}_k + \alpha_k \mathbf{d}_k, \tag{5.59}$$

where the step length α_k is selected appropriately to ensure the merit function,

$$\phi_1(\boldsymbol{\rho}; \alpha) = J(\boldsymbol{\rho}) + \alpha \|\mathbf{c}(\boldsymbol{\rho})\|_1, \tag{5.60}$$

satisfying the sufficient decrease condition.

The SQP iterations stop to reach the local minimizer when the first-order necessary condition is approached closely enough:

$$\left\| \begin{bmatrix} \nabla_{\mathbf{p}} \mathcal{L}(\boldsymbol{\rho}_k, \lambda_k) \\ \mathbf{c}(\boldsymbol{\rho}_k) \end{bmatrix} \right\|_2 < \varepsilon, \tag{5.61}$$

or until the maximum number of iterations is reached.

Take a typical lane change maneuver as an example, the optimized results of the curvature control parameterized as a spline for the problems (5.47)–(5.49) after SQP iterations are shown in Figure 5.29, as well as the optimized polynomial parameterization result for comparison. It is observed that the polynomial curvature retains its shape no matter how the weights in the utility functional (5.49) change. This is due to the searching scope for the curvature in a polynomial being strictly limited, while the curvature in splines is more versatile and is subject to change with respect to the form of utility functional.

On the other hand, solving the optimization via the control parameterization is computationally much more efficient than trying to solve the optimal control through a numerical approach such as using the dynamic programming (DP) algorithm. Using DP would require the formation

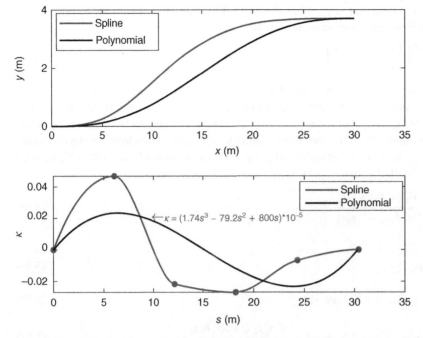

Figure 5.29 Optimized results in the spline or polynomial representation, respectively, for a typical lane-change maneuver. Source: Zhu and Aksun-Guvenc [57]/with permission from Springer Nature.

of a four-dimensional model with curvature rate as control in (5.40)–(5.42) to account for all state and control constraints. To get an accurate solution, a fine quantization to each state and control is required for DP. This high-dimensional problem associated with fine quantization forms the so-called "curse of dimensionality," which makes the DP approach rather impractical.

This versatility of the spline representation along with the flexible adjustment of the relative weight values in Eq. (5.49) could provide diverse feasible trajectories. Figure 5.30 shows three different sets of relative weights w_1, w_2 in (5.49) that mimic three different driving styles ranging from aggressive, normal, to gentle.

Since the trajectory optimization as described is a nonconvex nonlinear programming problem, the initial guess could affect the local convergence results and especially the convergence speed. A good initial guess of the curvature control spline parameters can effectively reduce the number of iterations required. Therefore, the lookup tables with optimal control parameter values can be precomputed and then applied online to provide interpolation results as initial guesses. The generation process of the lookup tables is illustrated in Figure 5.31 for one vehicle initial pose. Under the vehicle body-centered coordinate framework, the terminal states were sampled uniformly first before the corresponding optimal trajectories were generated. The control parameters for each sampled trajectory were then stored in 4D lookup tables. The curvature states were not used as indices to reduce the table dimensions.

Figure 5.32 uses one example to illustrate the reduction of iteration times during the SQP process after using the lookup table interpolation result as the initial guess for the curvature control parameters. The initial guess obtained from the lookup tables approximates the local minimizer quite close in Figure 5.32b.

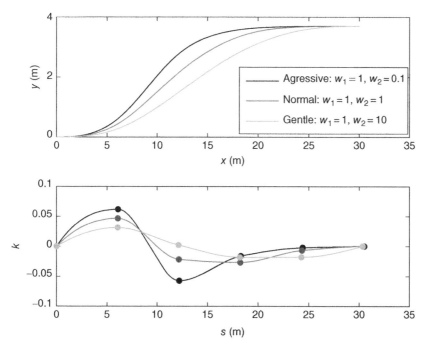

Figure 5.30 Different driving styles by adjusting relative weights in the utility functional. Source: Zhu and Aksun-Guvenc [57]/with permission from Springer Nature.

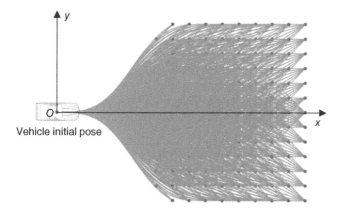

Figure 5.31 Visualization of the lookup table generation process at one initial pose. Source: Zhu and Aksun-Guvenc [57]/with permission from Springer Nature.

5.4.4 Sampling of the Longitudinal Movements

Since the vehicle motion models (5.40)–(5.42) decouple the velocity v, we are able to plan the longitudinal movement separately. A quintic polynomial is used to parameterize the longitudinal motion $s(t)$ as it is proven to be the jerk-optimal connection from a start state $S_0 = [s_0, \dot{s}_0, \ddot{s}_0]$ to an end state $S_1 = [s_1, \dot{s}_1, \ddot{s}_1]$ within the time interval $T := t_1 - t_0$ in a one-dimensional problem [33]. Moreover, the six coefficients of the quintic polynomial could be readily derived by substituting in the terminal states S_0 and S_1 as well as the time interval T.

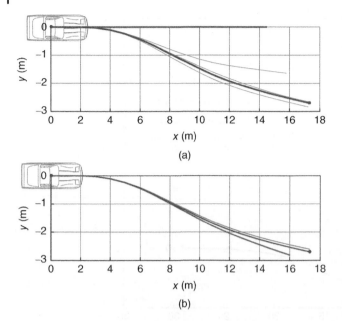

Figure 5.32 SQP iterations until convergence from the initial guess (a) before and (b) after applying the lookup tables. Source: Zhu and Aksun-Guvenc [57]/with permission from Springer Nature.

The start state S_0 is always known, while the end state S_1 and duration T could vary a lot depending on the desired vehicle behavior. In a structured road environment, the possible vehicle behaviors can be divided into four as follows: following, stopping, merging, and velocity keeping. We discuss the choice of S_1 and T for each behavior below.

Following, Stopping, or Merging

Under the following, stopping, or merging mode, the target trajectory describes the desired position $s_{target}(t)$ which possibly varies with time. It is to be decided when the current vehicle reaches into the target trajectory. Hence, we generate a set of longitudinal trajectories by varying the end constraints $[s_1, \dot{s}_1, \ddot{s}_1]$ with sampled time intervals T_j:

$$\left[s_1, \dot{s}_1, \ddot{s}_1, T\right]_j = \left[s_{target}(T_j), \dot{s}_{target}(T_j), \ddot{s}_{target}(T_j), T_j\right]. \tag{5.62}$$

The target trajectory $s_{target}(t)$ for the ego vehicle under the following mode could be derived by keeping a certain temporal safety distance from the leading vehicle ahead in the same lane:

$$s_{target}(t) = s_{lv}(t) - \left[D_0 + \tau \dot{s}_{lv}(t)\right], \tag{5.63}$$

where constants D_0 and τ adjust the standstill distance and constant time gap in between. $s_{lv}(t)$, $\dot{s}_{lv}(t)$, and $\ddot{s}_{lv}(t)$ are predicted longitudinal motion of the leading vehicle. The target velocity and acceleration at the following mode are simply the derivatives of (5.63):

$$\dot{s}_{target}(t) = \dot{s}_{lv}(t) - \tau \ddot{s}_{lv}(t), \tag{5.64}$$

$$\ddot{s}_{target}(t) = \ddot{s}_{lv}(t) - \tau \dddot{s}_{lv}(t) \approx \ddot{s}_{lv}(t). \tag{5.65}$$

For the merging mode, the ego vehicle tries to squeeze in between a pair of vehicles at $s_a(t)$ and $s_b(t)$ and the target position could be selected as follows:

$$s_{target}(t) = \frac{1}{2}\left[s_a(t) + s_b(t)\right], \tag{5.66}$$

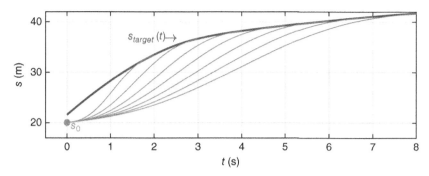

Figure 5.33 Sampling of jerk-optimal quintic trajectories to reach the target trajectory. Source: Zhu and Aksun-Guvenc [57]/with permission from Springer Nature.

with target velocity and acceleration obtained similarly with derivatives as in (5.64) and (5.65).

Under the stopping mode, the target position could be fixed depending on the intersection or road block $s_{target}(t) = s_{stop}$, $\dot{s}_{target}(t) = 0$, and $\ddot{s}_{target}(t) = 0$.

Figure 5.33 shows an example of sampling trajectories transitioning into the target trajectory. Not all the sampling ones are feasible for the vehicle to follow. We have yet to combine both the curvature and longitudinal trajectories and then evaluate their costs and feasibilities altogether.

Velocity Keeping

In situations other than the above, the ego vehicle does not necessarily have to be at a certain position at a specific time but needs to adapt to a desired speed $\dot{s}_d$. The desired speed is generally set as a constant right below the traffic speed limit but should also be constrained by a lateral acceleration threshold to ensure safety and comfort at large curvature such as during cornering. We achieve this by a preview of road curvature and take the $\dot{s}_d$ that obeys the speed limit and satisfies the lateral acceleration cap. Analogous to the proof with the calculus of variations in [60], a quartic polynomial minimizes the jerk functional for a given start state $S_0 = [s_0, \dot{s}_0, \ddot{s}_0]$ at t_0 and $[\dot{s}_1, \ddot{s}_1]$ of the end state S_1 at $t_1 = t_0 + T$. We can generate sampling speed trajectories by varying the time interval T_j according to

$$\left[\dot{s}_1, \ddot{s}_1, T\right]_j = \left[\dot{s}_d, 0, T_j\right], \tag{5.67}$$

The generated quartic can be converted to the quintic position trajectory form to maintain the uniformity with others generated under other behavior modes.

5.4.5 Trajectory Evaluation and Selection

Trajectory Candidates Generation

The generation methodology of the curvature and velocity (equivalently position) trajectories parameterized as in (5.43) is described in detail in Sections 5.4.2 and 5.4.3. The Frenet frame is used to form these trajectories to mimic the human maneuver to plan the vehicle's lateral movement relative to the lanes. As depicted in Figure 5.34, the Frenet frame is a moving reference frame formulated with the tangential and normal vectors $\vec{t}_r, \vec{n}_r$ at a certain point on the path. The Cartesian coordinate of the trajectory point of the center of mass of the vehicle can be related to its Frenet coordinates (s, d) or equivalently:

$$\vec{p}(s, d) = \vec{r}(s) + d \cdot \vec{n}_t, \tag{5.68}$$

as shown in Figure 5.34, where s denotes the arc length of the centerline.

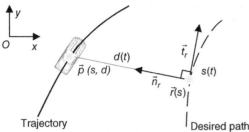

Figure 5.34 Trajectory representation in Cartesian and Frenet frame. Source: Zhu and Aksun-Guvenc [57]/with permission from Springer Nature.

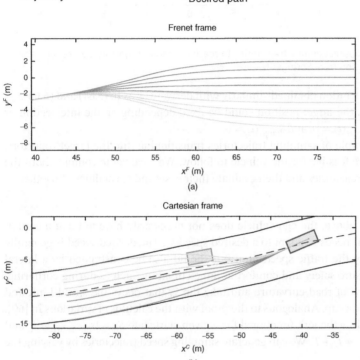

Figure 5.35 Trajectory candidates with optimized curvature controls in the Frenet frame (a), and their transformations back to Cartesian frame (b). Source: Zhu and Aksun-Guvenc [57]/with permission from Springer Nature.

To provide more reactive capabilities to the vehicle, a set of terminal states for each lane under the Frenet frame are chosen to generate the sampling trajectories in the configuration space with the parameterized curvature control as depicted in Figure 5.35. The planning horizon is set to vary with the speed but limited with a lower threshold. The trajectories in the configuration space are settled once the curvature controls $\kappa(\mathbf{p}, s)$ were optimized and are not subject to change with respect to the longitudinal movement, as revealed in the motion model (5.40)–(5.42). The sampling set of longitudinal movements is applied to each trajectory in Figure 5.35 to combine into a full set of sampling trajectories with both longitudinal and lateral movements integrated.

Trajectory Cost Evaluation and Collision Check

The combined trajectories are transformed back under the Cartesian coordinates (Figure 5.35) and then their costs are evaluated. The cost functional is defined as follows:

$$C = w_j \int_0^{t_f} \dddot{s}(t)dt + w_y \int_0^{t_f} a_y^2(t)dt + w_d \int_0^{t_f} d^2(t)dt + w_r \int_0^{t_f} (v(t) - v_r)^2 dt, \tag{5.69}$$

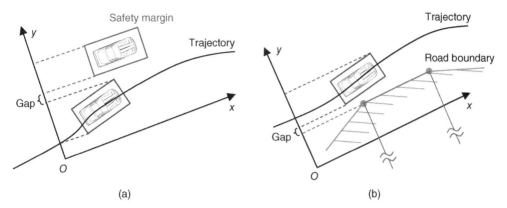

Figure 5.36 Application of the separating axis theorem to (a) collision check with ORU, or (b) violation into road boundaries. Source: Zhu and Aksun-Guvenc [57]/with permission from Springer Nature.

where the four integrals are associated with penalties to jerk, lateral acceleration a_y, offset d from the path, and deviation from the desired speed v_r. The trajectories with longitudinal, lateral, or combined acceleration exceeding the vehicle dynamic capabilities or the road friction limit are considered infeasible and are not used inside the cost functional.

The feasibility check for each trajectory also includes collision detection with ORU in the surrounding environment. To make the collision check computationally efficient, regular shapes such as circles or rectangles are often used to envelop the vehicle shape [13, 61, 62]. Here, we take the rectangle shapes with added safety margins to leave enough room between vehicles. The separating axis theorem can be applied to check if there is any overlap between two convex objects. It states that two convex objects do not overlap if there exists an axis onto which the projections of the two do not overlap. For the rectangles representation, this requires four projections at maximum with axes taken in parallel with rectangle sides for collision check (Figure 5.36).

This method can be expanded to check the vehicle's violation into the road or lane boundaries as well. As depicted in Figure 5.36b, each segment of the road boundary can be expanded into a rectangle with two sides extended long enough. Then, similar projection methods as used in collision check can be applied to examine road boundary violation.

The trajectories that pass the acceleration feasibility check, collision detection, and road boundary intrusion are considered feasible and safe. The one with the lowest cost functional values evaluated in (5.69) is selected for execution.

5.4.6 Integration of Road Friction Coefficient Estimation for Safety Enhancement

Path planning and path tracking are usually studied under the assumption of ideal dry road conditions. This assumption could pose potential dangers when driving in bad weather where water, snow, or ice forms at the road surface and significantly reduces the adhesion limit of the vehicle's tires. It is, hence, of great interest to bring the estimated knowledge of road friction coefficient into the trajectory planning and control framework to see if safety improvements can be achieved by their knowledge.

The slip slope method detailed in Chapter 3 is used for the estimation of road friction coefficient μ. The estimated μ implicitly confines the curvature control shape through the curvature constraints (5.44)–(5.46) for the model-based trajectory planning. Figure 5.37 shows constrained optimization results of trajectory and corresponding curvature control for different μ values at the

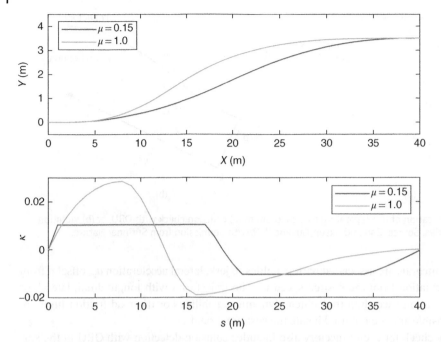

Figure 5.37 Constrained optimization results of trajectory and curvature control for different μ at speed 10 m/s. Source: Zhu [63].

speed of 10 m/s. By restraining the curvature with respect to the value of μ, the planned trajectory at low μ is gentler than the one at high μ as shown in Figure 5.37.

Note that the estimated value of μ will continuously feed to the trajectory planning module to generate new trajectories from new vehicle states to potential terminal states. The vehicle will select the safe one with minimum cost of Eq. (5.69) out of all candidate trajectories. The selected one is then followed by a PID-tuned longitudinal control and the MPC path following controller detailed in Chapter 7.

A lane-changing scenario is designed in simulation for evaluating the case where the ego vehicle has to avoid a static object on an icy road with μ set to 0.15. Figure 5.38 shows the comparison of vehicle trajectories with and without the μ estimation module, as well as the continuously planned trajectory. The vehicle starts at 7.5 m/s initial speed with a target speed of 10 m/s. The estimation module is capable of identifying the road friction coefficient close to reality through the slip slope method during the required acceleration period as shown in Figure 5.39. The planning with the

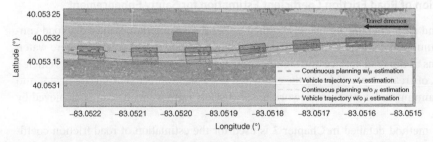

Figure 5.38 Continuously planned and vehicle actual trajectories at icy road $\mu = 0.15$ with and without the road friction coefficient estimation module. Source: Zhu [63].

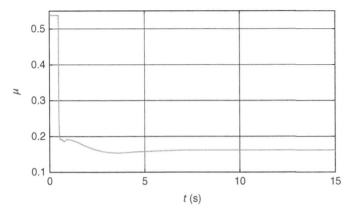

Figure 5.39 Estimated road friction coefficient fed to the trajectory planning module. Source: Zhu [63].

knowledge of the continuously updated μ value leads to a gentler trajectory in Figure 5.38, which is dynamically feasible for the vehicle to follow. On the contrary, the trajectory that was designed assuming ideal road condition $\mu = 1$ without the μ estimation module appears too aggressive for such icy road. As a result, the vehicle actually follows the trajectory in light gray solid line which takes the vehicle partially off the road due to saturation of the front steering tires.

Stroboscopic views of dynamic feasible candidate trajectories and the selected one in Figure 5.40 reveals the difference of trajectory planning under the two cases. In the plot, the multiple, open ended trajectories at each moment are potential candidates that are considered to be dynamically feasible by the planning module. With the μ estimation module being available, scarcer candidates are considered feasible under the low friction road, compared to the planned trajectories with

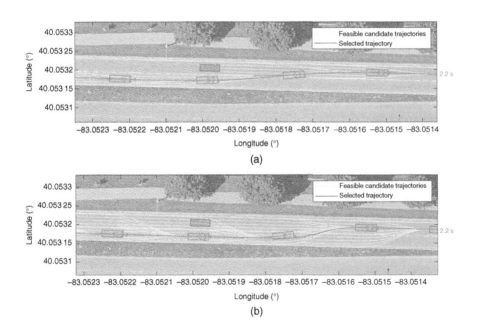

Figure 5.40 Stroboscopic views of the selected trajectory among the dynamically feasible candidates for the two cases in Figure 5.38: (a) with estimated μ; (b) with incorrect nominal μ. Source: Zhu [63].

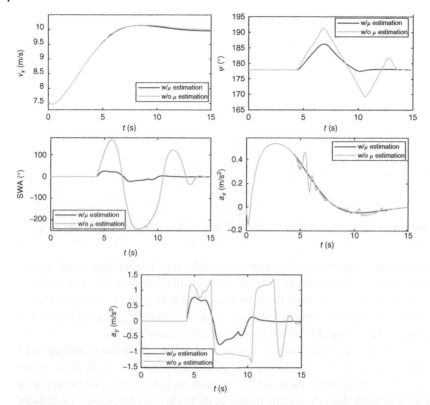

Figure 5.41 Comparison of vehicle states of speed, heading, steering wheel angle, longitudinal, and lateral acceleration, respectively, for the two cases in Figure 5.38 with and without the μ estimation module. Source: Zhu [63].

nominal μ. The candidate trajectories under the estimated μ are also gentler in contrast, similar to what was revealed in Figure 5.37.

Comparisons of vehicle states such as longitudinal speed, vehicle heading, steering wheel angle, longitudinal, and lateral acceleration under the two cases are also made as shown in Figure 5.41. The gentle trajectory planning based on estimated μ led to reasonable low-level control response, with smaller and gentler steering, and accelerations within the road adhesion limit. Meanwhile, the response under the other trajectory without the μ estimation module is not favorable. It appears that the vehicle front tires reach the road adhesion limit quickly and do not provide sufficient lateral force for accurate trajectory tracking. This, in turn, leads to steering overshoot by the MPC steering controller used and poses a risk for potential loss of vehicle stability.

It is demonstrated that the integration of the road friction coefficient estimation module into the trajectory planning and control framework has a great potential to improve safety of autonomous driving with the knowledge of the estimated μ. The trajectory planning method raised in this section is capable of adjusting the trajectory to satisfy the control constraints under this estimated road friction coefficient.

5.4.7 Simulation Results in Complex Scenarios

The proposed trajectory planning framework was implemented in a real-time simulated environment with Simulink Real-Time™ kernel using a PC with 2.60 GHz CPU. A high-fidelity CarSim

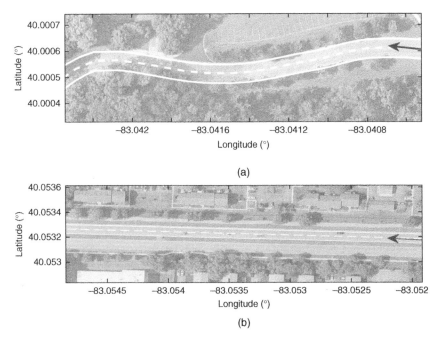

Figure 5.42 Two road segments selected for evaluation: a curved road with 25 mph speed limit (a), and a straight road with 45 mph speed limit (b). Source: Zhu and Aksun-Guvenc [57]/with permission from Springer Nature.

vehicle model with parameters matching our full-scale experimental vehicle was used. Two road segments were selected as shown in Figure 5.42 to evaluate the trajectory planning and following framework in dynamic scenarios with one or more surrounding vehicles: one curved road with 25 mph speed limit to also observe the speed adaptivity to curvature change, and one straight road with 45 mph speed limit to observe the performance at higher speed. The trajectory planning runs with a cycle time of 100 ms, while the low-level longitudinal and lateral controls run every 10 ms. Note that predicted movements of other vehicles were assumed given in the short preview horizon (three seconds), which usually rely on perception sensors or vehicular network and form another research topic. Scenario descriptions and results are presented in the following.

Velocity Keeping Mode Only
In these test scenarios, we switched off the behavior layer and let the vehicle run in velocity keeping mode only. This is to observe the capabilities of the reactive layer in terms of collision avoidance and adaptability to large curvature road.

Figure 5.43 shows the continuous movements of the ego vehicle and its surrounding vehicles. The stroboscopic shapes in every 0.3 seconds depict the planned/predicted motion trajectories of the vehicles for the future three seconds. The bottom four plots in Figure 5.43 are time plots of the vehicle speed, lateral offset from the right lane centerline (ordinate y^F of the Frenet frame), steering wheel angle, and longitudinal/lateral accelerations. The vertical and gray colored, dashed lines in these plots are to indicate the important instances of the motion diagrams above them in Figure 5.43.

The vehicle travels with an initial speed of 2.8 m/s and starts to accelerate to the desired speed capped by the speed limit. At $t = 5.8$ seconds, the ego vehicle encounters the first slow traveling vehicle ahead and plans a trajectory right beside it to overtake. Then, at $t = 7.7$ seconds, the ego

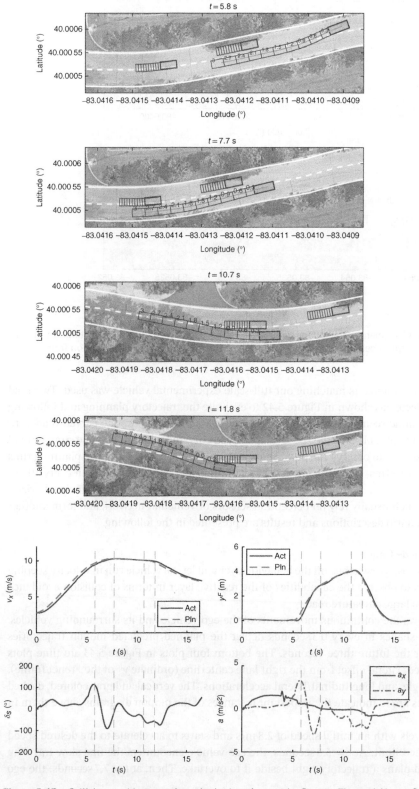

Figure 5.43 Collision avoidance under velocity keeping mode. Source: Zhu and Aksun-Guvenc [57]/with permission from Springer Nature.

vehicle has to plan a trajectory with larger lateral offset to avoid the second slow vehicle ahead near the road center. This explains the increase of lateral offset y^F. The diagrams at $t = 10.7$ seconds and $t = 11.8$ seconds show the process when the vehicle chooses the trajectory back to the right lane.

The planned velocity and y^F curves are also shown in the time plots of Figure 5.43 for comparison with the actual vehicle states. It is apparent that the actual velocity and lateral movement track the planned ones quite accurately, indicating that both the longitudinal and lateral controller worked well.

It can also be seen that the vehicle slows down when it approaches the large curvature at the end of this road segment (Figure 5.42). This speed adaptivity is due to the fact that terminal speed at every planning instance is limited by both the speed limit and the largest curvature observed over the preview horizon.

It needs to be pointed out that even though the trajectories planned with the reactive layer only are feasible and safe without collisions, they may not be desirable by the driver. For example, at $t = 10.7$ seconds and $t = 11.8$ seconds, the ego vehicle chose to cut back to the right lane immediately after overtaking, which was considered quite aggressive without leaving some room with the overtaken vehicle behind. This makes an upper behavior layer necessary to decide the mode (velocity keeping, following/stopping, or merging) for the reactive layer to carry on.

Lane Change Due to Road Blockade

In the following two scenarios, a simplified rule-based behavior layer is formed to decide the mode for the reactive layer to plan corresponding motions. In this scenario, as shown in Figure 5.44, a roadblock was placed on the lane where the ego vehicle travels at an initial speed of 10 m/s. The text at the right bottom corner of each motion diagram in Figure 5.44 indicates the mode set by the behavior layer ("VK" for velocity keeping, "FL" for following/stopping, and "MG" for merging). At $t = 1.3$ seconds, the ego vehicle slows down until it makes a full stop under the following/stopping mode as the other lane is not safe for lane changing due to the existence of another vehicle. Then at $t = 5.9$ seconds, the vehicle switches from the following/stopping mode to velocity keeping mode, and starts to initiate a lane-change maneuver. The ego vehicle then follows the vehicle ahead ($t = 9.8$ seconds) until it surpasses the roadblock. Once the right lane is clear, the ego vehicle switches back to the velocity-keeping mode again and plans a trajectory to the right lane at its faster desired speed ($t = 12.2$ seconds).

Merging Between Cars

The third scenario shows how the vehicle merges in traffic at a higher speed in the straight road segment (Figure 5.45). The vehicle travels initially at 20 m/s but has to slow down due to a slow vehicle in front ($t = 1.2$ seconds). The ego vehicle keeps following until it finds a gap between the two vehicles on the other lane, and plans a merging motion at $t = 3.8$ seconds. Note how the vehicle accelerates its speed in order to merge into the faster traffic within the 3.8–7.3 seconds time interval. Once the vehicle is finished with merging, it switches to the following mode to maintain some distance (dependent on the front vehicle speed) from the vehicle ahead ($t = 7.3$ seconds). Then at $t = 14.7$ seconds, the vehicle initiates lane change to approach closer to its desired speed of 20 m/s again. The completion of this merging scenario in real-time simulation shows that the proposed planning framework has potential to be applied at high speed as well. One thing that needs to be mentioned is that the strategy may be limited by its planning cycle, between which the vehicle is not responsive to an immediate threat. For example, with the current planning cycle at 10 Hz, the vehicle at 30 m/s speed will already travel 3 m before its next planning instance, leaving it vulnerable to emergent conditions. This planning frequency may be increased by using more powerful computational hardware or simplified optimization programming techniques in the future.

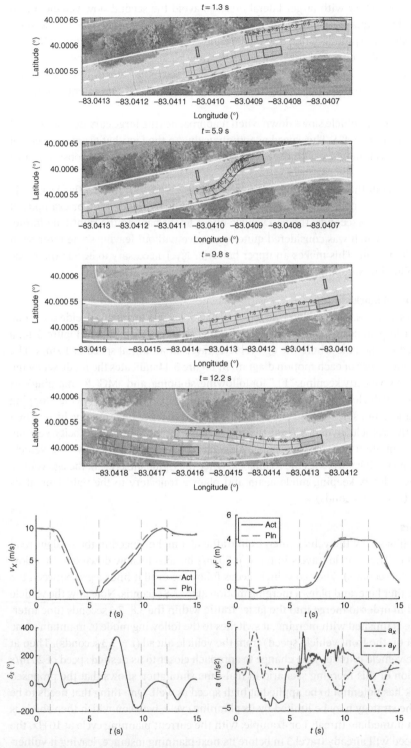

Figure 5.44 Wait for lane change due to roadblock ahead. Source: Zhu and Aksun-Guvenc [57]/with permission from Springer Nature.

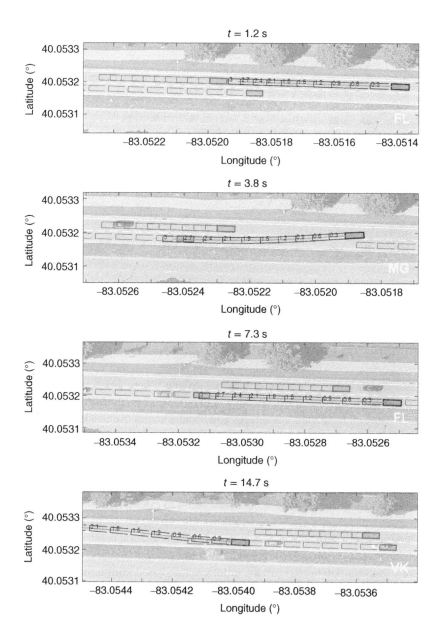

Figure 5.45 Merging into traffic. Source: Zhu and Aksun-Guvenc [57] (© [2020]) with permission from Springer Nature.

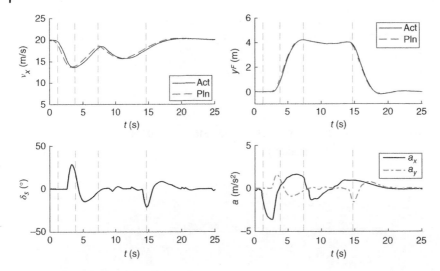

Figure 5.45 (*Continued*)

5.5 Chapter Summary and Concluding Remarks

Collision avoidance has a significant role in safer automated driving. There are a wide variety of approaches to this problem in the literature. In this chapter, several methods for collision avoidance were discussed. They all have their advantages and disadvantages over each other and can be used alternatively depending on the application. Each method was presented in detail and formulated accordingly. Results were provided for model in the loop (MIL), HIL simulation tests, and/or real-world experiments.

First of all, the elastic band method was introduced and discussed. This method is based on considering the path as an elastic band and modifying it according to external forces created by obstacles and internal forces which nodes apply on each other to keep the elastic band together. This method results in a very smooth path which reduces the lateral acceleration of the vehicle and hence provides comfortable maneuvering. Also, it has relatively low computational cost. Second, as an alternative with very low computational cost, the quintic spline-based minimum curvature variation method was presented and discussed. In this method, the path generation process was enhanced by using previously generated optimal parameter lookup tables to significantly reduce the computation time. The resulting path was spline based. Third, method that also takes the vehicle dynamic constraints and longitudinal motion into account was presented and discussed. Although it has higher computational cost with respect to the other methods, the model-based trajectory planning method offers a more analytical and comprehensive solution.

References

1 M. W. Mueller, M. Hehn, and R. D'Andrea, "A computationally efficient motion primitive for quadrocopter trajectory generation," *IEEE Transactions on Robotics*, vol. 31, no. 6, pp. 1294–1310, 2015.

2 M. Vukosavljev, Z. Kroeze, A. P. Schoellig, and M. E. Broucke, "A modular framework for motion planning using safe-by-design motion primitives," *IEEE Transactions on Robotics,* vol. 35, no. 5, pp. 1233–1252, 2019.

3 J. Alonso-Mora, P. Beardsley, and R. Siegwart, "Cooperative collision avoidance for nonholonomic robots," *IEEE Transactions on Robotics,* vol. 34, no. 2, pp. 404–420, 2018.

4 F. Mohseni and L. Nielsen, "Decoupled sampling-based velocity tuning and motion planning method for multiple autonomous vehicles," *2018 IEEE Intelligent Vehicles Symposium (IV), Changshu, Suzhou, China,* 2018, pp. 1–6: IEEE, doi:https://doi.org/10.1109/IVS.2018.8500619.

5 C.-F. Lin, J.-C. Juang, and K.-R. Li, "Active collision avoidance system for steering control of autonomous vehicles," *IET Intelligent Transport Systems,* vol. 8, no. 6, pp. 550–557, 2014.

6 S. Dixit, U. Montanaro, S. Fallah, *et al.,* "Trajectory planning for autonomous high-speed overtaking using MPC with terminal set constraints," *2018 21st International Conference on Intelligent Transportation Systems (ITSC),* 2018, pp. 1061–1068: IEEE, doi:https://doi.org/10 .1109/ITSC.2018.8569529.

7 S. Quinlan and O. Khatib, "Elastic bands: connecting path planning and control," in *Proceedings IEEE International Conference on Robotics and Automation,* 1993, pp. 802–807: IEEE.

8 O. Khatib, "Real-time obstacle avoidance for manipulators and mobile robots," in *IEEE International Conference on Robotics and Automation,* 1985, vol. 2, pp. 500–505: IEEE.

9 T. Sattel and T. Brandt, "Ground vehicle guidance along collision-free trajectories using elastic bands," in *American Control Conference,* 2005, pp. 4991–4996: IEEE.

10 S. K. Gehrig and F. J. Stein, "Collision avoidance for vehicle-following systems," *IEEE Transactions on Intelligent Transportation Systems,* vol. 8, no. 2, pp. 233–244, 2007.

11 Ö. Ararat and B. A. Güvenç, "Development of a collision avoidance algorithm using elastic band theory," *IFAC Proceedings Volumes,* vol. 41, no. 2, pp. 8520–8525, 2008.

12 M. T. Emirler, H. Wang, and B. A. Güvenç, "Socially acceptable collision avoidance system for vulnerable road users," *IFAC-PapersOnLine,* vol. 49, no. 3, pp. 436–441, 2016.

13 J. Ziegler, P. Bender, T. Dang, and C. Stiller, "Trajectory planning for Bertha – a local, continuous method," in *2014 IEEE Intelligent Vehicles Symposium Proceedings,* 2014, pp. 450–457: IEEE.

14 T. Gu and J. M. Dolan, "On-road motion planning for autonomous vehicles," in *Intelligent Robotics and Applications,* Berlin, Heidelberg, 2012, pp. 588–597: Springer Berlin Heidelberg.

15 X. Li, Z. Sun, D. Cao, D. Liu, and H. He, "Development of a new integrated local trajectory planning and tracking control framework for autonomous ground vehicles," *Mechanical Systems and Signal Processing,* vol. 87, no. Part B, pp. 118–137, 2017.

16 D. Gonzalez, J. Perez, V. Milanes, and F. Nashashibi, "A review of motion planning techniques for automated vehicles," *IEEE Transactions on Intelligent Transportation Systems,* vol. 17, no. 4, pp. 1135–1145, 2016.

17 L. Ma, J. Xue, K. Kawabata, J. Zhu, C. Ma, and N. Zheng, "Efficient sampling-based motion planning for on-road autonomous driving," *IEEE Transactions on Intelligent Transportation Systems,* vol. 16, no. 4, pp. 1961–1976, 2015.

18 S. Karaman and E. Frazzoli, "Sampling-based algorithms for optimal motion planning," *International Journal of Robotics Research,* vol. 30, no. 7, pp. 846–894, 2011.

19 S. Yoon, D. Lee, J. Jung, and D. H. Shim, "Spline-based RRT* using piecewise continuous collision-checking algorithm for car-like vehicles," *Journal of Intelligent and Robotic Systems,* vol. 90, no. 3–4, pp. 537–549, 2017.

20 S. Kato, E. Takeuchi, Y. Ishiguro, Y. Ninomiya, K. Takeda, and T. J. I. M. Hamada, "An open approach to autonomous vehicles," *IEEE Micro*, vol. 35, no. 6, pp. 60–68, 2015, doi:https://doi .org/10.1109/MM.2015.133.

21 C. Zhang, D. Chu, S. Liu, Z. Deng, C. Wu, and X. Su, "Trajectory planning and tracking for autonomous vehicle based on state lattice and model predictive control," *IEEE Intelligent Transportation Systems Magazine*, vol. 11, no. 2, pp. 29–40, 2019.

22 D. Dolgov and S. Thrun, "Autonomous driving in semi-structured environments: mapping and planning," in *2009 IEEE International Conference on Robotics and Automation*, 2009, pp. 3407–3414: IEEE.

23 S. Sedighi, D.-V. Nguyen, and K.-D. Kuhnert, "Guided hybrid A-star path planning algorithm for valet parking applications," in *2019 5th International Conference on Control, Automation and Robotics (ICCAR)*, 2019, pp. 570–575: IEEE.

24 J. Ji, A. Khajepour, W. W. Melek, and Y. Huang, "Path planning and tracking for vehicle collision avoidance based on model predictive control with multiconstraints," *IEEE Transactions on Vehicular Technology*, vol. 66, no. 2, pp. 952–964, 2017.

25 O. Montiel, U. Orozco-Rosas, and R. Sepúlveda, "Path planning for mobile robots using Bacterial Potential Field for avoiding static and dynamic obstacles," *Expert Systems with Applications*, vol. 42, no. 12, pp. 5177–5191, 2015.

26 H. Guo, C. Shen, H. Zhang, H. Chen, and R. Jia, "Simultaneous trajectory planning and tracking using an MPC method for cyber-physical systems: a case study of obstacle avoidance for an intelligent vehicle," *IEEE Transactions on Industrial Informatics*, vol. 14, no. 9, pp. 4273–4283, 2018.

27 B. Yi, S. Gottschling, J. Ferdinand, N. Simm, F. Bonarens, and C. Stiller, "Real time integrated vehicle dynamics control and trajectory planning with MPC for critical maneuvers," in *2016 IEEE Intelligent Vehicles Symposium (IV)*, 2016, pp. 584–589: IEEE.

28 S. Y. Gelbal, B. Aksun-Guvenc, and L. Guvenc, "Collision avoidance of low speed autonomous shuttles with pedestrians," *International Journal of Automotive Technology*, vol. 21, no. 4, pp. 903–917, 2020.

29 Q. H. Do, L. Han, H. T. N. Nejad, and S. Mita, "Safe path planning among multi obstacles," in *2011 IEEE Intelligent Vehicles Symposium (IV)*, 2011, pp. 332–338: IEEE.

30 K. Osman, J. Ghommam, and M. Saad, "Guidance based lane-changing control in high-speed vehicle for the overtaking maneuver," *Journal of Intelligent and Robotic Systems*, vol. 98, pp. 643–665, 2020, doi:https://doi.org/10.1007/s10846-019-01070-6.

31 F. You, R. Zhang, G. Lie, H. Wang, H. Wen, and J. Xu, "Trajectory planning and tracking control for autonomous lane change maneuver based on the cooperative vehicle infrastructure system," *Expert Systems with Applications*, vol. 42, no. 14, pp. 5932–5946, 2015.

32 I. A. Ntousakis, I. K. Nikolos, and M. Papageorgiou, "Optimal vehicle trajectory planning in the context of cooperative merging on highways," *Transportation Research Part C Emerging Technologies*, vol. 71, pp. 464–488, 2016.

33 M. Werling, S. Kammel, J. Ziegler, and L. Gröll, "Optimal trajectories for time-critical street scenarios using discretized terminal manifolds," *International Journal of Robotics Research*, vol. 31, no. 3, pp. 346–359, 2011.

34 X. Hu, L. Chen, B. Tang, D. Cao, and H. He, "Dynamic path planning for autonomous driving on various roads with avoidance of static and moving obstacles," *Mechanical Systems and Signal Processing*, vol. 100, pp. 482–500, 2018.

35 L. Ma, J. Yang, and M. Zhang, "A two-level path planning method for on-road autonomous driving," in *2012 Second International Conference on Intelligent System Design and Engineering Application*, 2012, pp. 661–664: IEEE.

36 N. Nagasaka and M. Harada, "Towards safe, smooth, and stable path planning for on-road autonomous driving under uncertainty," in *2016 IEEE 19th International Conference on Intelligent Transportation Systems (ITSC)*, 2016, pp. 795–801: IEEE.

37 T. M. Howard and A. Kelly, "Optimal rough terrain trajectory generation for wheeled mobile robots," *International Journal of Robotics Research*, vol. 26, no. 2, pp. 141–166, 2007.

38 J. Ziegler, P. Bender, M. Schreiber *et al.*, "Making Bertha drive – an autonomous journey on a historic route," *IEEE Intelligent Transportation Systems Magazine*, vol. 6, no. 2, pp. 8–20, 2014, doi:https://doi.org/10.1109/MITS.2014.2306552.

39 E. Swanson, M. Yanagisawa, W. Najm, F. Foderaro, and P. Azeredo, "Crash avoidance needs and countermeasure profiles for safety applications based on light-vehicle-to-pedestrian communications," National Highway Traffic Safety Administration, United States, 2016.

40 S. Dent, 2019, "Uber self-driving car involved in fatal crash couldn't detect jaywalkers," Available: https://www.engadget.com/2019/11/06/uber-self-driving-car-fatal-accident-ntsb/.

41 S. Zhu, S. Y. Gelbal, and B. Aksun-Guvenc, "Online Quintic Path Planning of Minimum Curvature Variation with Application in Collision Avoidance," *2018 21st International Conference on Intelligent Transportation Systems (ITSC)*, 2018, pp. 669–676: IEEE.

42 C. G. L. Bianco and A. Piazzi, "Optimal trajectory planning with quintic G/sup 2/-splines," in *Proceedings of the IEEE Intelligent Vehicles Symposium*, 2000, pp. 620–625: IEEE.

43 C. Katrakazas, M. Quddus, W.-H. Chen, and L. Deka, "Real-time motion planning methods for autonomous on-road driving: state-of-the-art and future research directions," *Transportation Research Part C Emerging Technologies*, vol. 60, pp. 416–442, 2015.

44 L. B. Cremean, T. B. Foote, J. H. Gillula, *et al.*, "Alice: an information-rich autonomous vehicle for high-speed desert navigation," *Journal of Field Robotics*, vol. 23, no. 9, pp. 777–810, 2006.

45 D. Dolgov, S. Thrun, M. Montemerlo, and J. Diebel, "Path planning for autonomous vehicles in unknown semi-structured environments," *International Journal of Robotics Research*, vol. 29, no. 5, pp. 485–501, 2010.

46 S. Thrun, M. Montemerlo, H. Dahlkamp, *et al.*, "Stanley: the robot that won the DARPA Grand Challenge," *Journal of Field Robotics*, vol. 23, no. 9, pp. 661–692, 2006.

47 D. Ferguson, T. M. Howard, and M. Likhachev, "Motion planning in urban environments," *Journal of Field Robotics*, vol. 25, no. 11–12, pp. 939–960, 2008.

48 H. Wang, A. Tota, B. Aksun-Guvenc, and L. Guvenc, "Real time implementation of socially acceptable collision avoidance of a low speed autonomous shuttle using the elastic band method," *Mechatronics*, vol. 50, pp. 341–355, 2018.

49 L. Tang, S. Dian, G. Gu, K. Zhou, S. Wang, and X. Feng, "A novel potential field method for obstacle avoidance and path planning of mobile robot," in *3rd International Conference on Computer Science and Information Technology*, 2010, vol. 9, pp. 633–637: IEEE.

50 J. Levinson, J. Askeland, J. Becker, *et al.*, "Towards fully autonomous driving: systems and algorithms," in *IEEE Intelligent Vehicles Symposium (IV)*, 2011, pp. 163–168: IEEE.

51 T. Luettel, M. Himmelsbach, and H.-J. Wuensche, "Autonomous ground vehicles – concepts and a path to the future," *Proceedings of the IEEE*, vol. 100, no. Special Centennial Issue, pp. 1831–1839, 2012.

52 L. Han, H. Yashiro, H. T. N. Nejad, Q. H. Do, and S. Mita, "Bezier curve based path planning for autonomous vehicle in urban environment," in *2010 IEEE Intelligent Vehicles Symposium*, 2010, pp. 1036–1042: IEEE.

53 C. Urmson, J. Anhalt, D. Bagnell, *et al.*, "Autonomous driving in urban environments: Boss and the Urban Challenge," *Journal of Field Robotics*, vol. 25, no. 8, pp. 425–466, 2008.

54 F. Altché, P. Polack, and A. de La Fortelle, "High-speed trajectory planning for autonomous vehicles using a simple dynamic model," in *2017 IEEE 20th International Conference on Intelligent Transportation Systems (ITSC)*, 2017, pp. 1–7: IEEE.

55 A. Kelly and B. Nagy, "Reactive nonholonomic trajectory generation via parametric optimal control," *International Journal of Robotics Research*, vol. 22, no. 7–8, pp. 583–601, 2003.

56 J. Nocedal and S. Wright, *Numerical Optimization*, Springer Science & Business Media, 2006.

57 S. Zhu, and B. Aksun-Guvenc, "Trajectory planning of autonomous vehicles based on parameterized control optimization in dynamic on-road environments," *Journal of Intelligent & Robotic Systems*, vol. 100, pp. 1055–1067, 2020.

58 Q. H. Do, H. Tehrani, S. Mita, M. Egawa, K. Muto, and K. Yoneda, "Human drivers based active-passive model for automated lane change," *IEEE Intelligent Transportation Systems Magazine*, vol. 9, no. 1, pp. 42–56, 2017.

59 S. Zhu, S. Y. Gelbal, B. Aksun-Guvenc, and L. Guvenc, "Parameter-space based robust gain-scheduling design of automated vehicle lateral control," *IEEE Transactions on Vehicular Technology*, vol. 68, no. 10, pp. 9660–9671, 2019.

60 A. Talcahashi, T. Hongo, Y. Ninomiya, and G. Sugimoto, "Local path planning and motion control for agv in positioning," in *Proceedings. IEEE/RSJ International Workshop on Intelligent Robots and Systems' (IROS '89) The Autonomous Mobile Robots and Its Applications*, 1989, pp. 392–397, doi:https://doi.org/10.1109/IROS.1989.637936.

61 J. Ziegler and C. Stiller, "Fast collision checking for intelligent vehicle motion planning," in *2010 IEEE Intelligent Vehicles Symposium*, 2010, pp. 518–522: IEEE.

62 W. Lim, S. Lee, M. Sunwoo, and K. Jo, "Hierarchical trajectory planning of an autonomous car based on the integration of a sampling and an optimization method," *IEEE Transactions on Intelligent Transportation Systems*, vol. 19, no. 2, pp. 613–626, 2018.

63 S. Zhu, "Path Planning and Robust Control of Autonomous Vehicles," *Doctoral dissertation*, Ohio State University, 2020.

6

Path-Tracking Model Regulation

A model regulation approach is introduced in this chapter to handle uncertainty, disturbances, and time delay in the path-tracking model. This model regulation, also called the disturbance observer (DOB), is first applied to the loop from steering input to path-tracking error and subsequently used within a feedback control loop. It is seen to be an excellent path curvature rejection filter that also provides regulation of the plant around its nominal or desired model. This architecture also has inherent yaw stability control characteristics, thus, eliminating the need for an extra electronic stability control unit. Its time delay compensation capabilities and limitations are presented for the case of steering actuation delay due to the use of the controller area network (CAN) bus for steering actuation in autonomous vehicles.

6.1 Introduction

The models we create for representing real-world systems inherently contain different types of modeling errors. No matter how complicated and detailed the modeling is or how good the parameters are estimated using well-designed experiments, achieving a validated model that perfectly matches the actual physical plant is not possible. In order to achieve insensitivity against modeling errors due to model reduction, linearization, and other causes of model uncertainties, the model regulation architecture can be used [1]. The model regulator is also called the disturbance observer (DOB) as it has very good disturbance rejection properties [2]. The model regulation DOB loop partly overcomes the modeling mismatch and forces the system to behave like its desired or nominal model. The DOB structure is a two degrees-of-freedom (2 DOF) control structure. The DOB manipulates the response of the real-world system toward the response of its desired model representation by modifying the input. It was introduced by Ohnishi [3] and further refined by Umeno and Hori [4]. It has been successfully applied in a variety of applications. Some of these applications are friction compensation [5], road vehicle yaw stability control [6, 7], robust atomic force microscope control [8], power-assisted electric bicycle control [9], table drive system [10], and hard-disc-drive servo system [11]. The DOB architecture has excellent disturbance rejection properties, which makes it very useful for automated driving applications. This is due to the fact that the road curvature enters the path-tracking control loop as a disturbance. When used in this context, the DOB becomes a Road Curvature Rejection Filter.

Derived from the main DOB structure as its extensions, there are the communication disturbance observer (CDOB) and double disturbance observer (DDOB) structures. CDOB [12] allows us to estimate and compensate the time delay within the system, independent from the amount of delay. This allows compensation of time delay even in systems with variable time delay. Unlike

Autonomous Road Vehicle Path Planning and Tracking Control, First Edition.
Levent Güvenç, Bilin Aksun-Güvenç, Sheng Zhu, and Şükrü Yaren Gelbal.

DOB, however, CDOB has no disturbance rejection properties. In order to combine the disturbance rejection properties of DOB and the time delay compensation of CDOB, the DDOB structure was introduced in [13]. The DDOB is achieved by using both of the CDOB and DOB loops in the right manner. CDOB and DDOB are not used in this chapter as they are developed for a feedback loop with a reference input, i.e. command following, which is not the case for the path-tracking model that is used in this book where there is no reference input.

In this chapter, these control structures and how to apply them into several case studies related to automated driving are discussed. There are several application examples for different application cases, all involving path-tracking. Disturbance rejection analysis and Q filter design guidelines are also presented.

6.2 DOB Design and Frequency Response Analysis

The augmentation of a plant with the DOB forces it to behave just like its nominal (or desired) model within the bandwidth of the DOB. This is called model regulation. As DOB is used for model regulation, its structure has to be built around the plant model with uncertainty and disturbance in simulations and around the actual plant in experiments. The DOB also rejects disturbances within its bandwidth of operation and is, therefore, commonly used in disturbance rejection applications. The derivation of the model regulation using disturbance observer, analysis, and comparison of frequency response between the controlled and regulated systems are presented here for four different applications.

6.2.1 DOB Derivation and Loop Structure

The reader should recall Eq. (2.97) (repeated below) that relates the steering wheel angle input δ_f to the preview length projected path following error e_y output

$$e_y = G_{y\delta}(s, V)\delta_f + G_{y\rho}(s, V)\rho_{ref} + G_{yM}(s, V)M_{zd} \tag{6.1}$$

where the road curvature ρ_{ref} and external yaw moments M_{zd} are disturbances to be rejected. This is illustrated in Figure 6.1.

Consider multiplicative model uncertainty Δ_m in the nominal path following model G_n in Figure 6.1. Note that connections into junction points in the block diagrams in this chapter will not carry any sign if they are additive, but they will carry a minus $(-)$ sign if they are subtractive. Equation (6.1) is rewritten using the nominal model with multiplicative uncertainty, adding sensor noise n to path following error e_y and dropping the arguments as

$$e_y + n = G_n(1 + \Delta_m)\delta_f + G_{y\rho}\rho_{ref} + G_{yM}M_{zd}. \tag{6.2}$$

The model regulation aim in DOB design is to obtain the output that would normally be expected from the nominal plant, by using a slightly modified new input. We can write this goal condition as

$$e_y = G_n\delta_{fn}, \tag{6.3}$$

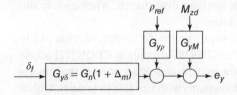

Figure 6.1 Path-tracking plant with steering input and disturbances.

which is the desired and expected input–output relation in the absence of model uncertainty, external disturbance, and sensor noise. δ_{fn} is a new steering wheel input signal to be defined below. Lumping together and treating all the curvature and yaw moment disturbances and model uncertainty in Eq. (6.2) as an extended disturbance d and solving for it results in

$$e_y + n = G_n\delta_f + (G_n\Delta_m\delta_f + G_{y\rho}\rho_{ref} + G_{yM}M_{zd}) = G_n\delta_f + d, \tag{6.4}$$

$$d = (e_y + n) - G_n\delta_f. \tag{6.5}$$

Then, if we define the control law

$$\delta_f = \delta_{fn} - \frac{d}{G_n}, \tag{6.6}$$

for δ_f in the right-hand side of Eq. (6.4), we obtain our desired input–output goal of Eq. (6.3) (except for the sensor noise) which includes both goals of regulation to the desired model G_n and rejection of the curvature and yaw moment disturbances. By substituting for the extended disturbance d from Eq. (6.5) into (6.6), we obtain the new version of the control law given by

$$\delta_f = \delta_{fn} - \frac{1}{G_n}(e_y + n) + \delta_f. \tag{6.7}$$

The control law in Eq. (6.7) works perfectly when substituted, but it cannot be implemented as we have the same steering wheel angle input on both sides of the equation. This is not causal and hence not implementable. To circumvent this problem and make Eq. (6.7) implementable, we multiply all the feedback signals on the right-hand side with the filter Q to obtain

$$\delta_f = \delta_{fn} - \frac{Q}{G_n}(e_y + n) + Q\delta_f. \tag{6.8}$$

Q is chosen as a low pass filter with a d.c. gain of unity. A simple interpretation is that $Q = 1$ is the ideal case that happens at low frequencies and corresponds to Eq. (6.7) which will exactly cancel the effect of the extended disturbance d when substituted back. $Q = 0$ which is the situation at high frequencies means that the DOB compensation is switched off. Limiting the DOB compensation to a preselected low-frequency range using the Q filter makes DOB implementable, avoids overcompensation at high frequencies such that it does not respond to high-frequency noise unnecessarily and also avoids stability robustness problems due to unmodeled uncertainty at high frequencies.

Using the control law in Eq. (6.8), we can create the DOB loop for path-tracking as shown in Figure 6.2.

As a result, the loop gain L of the DOB compensated path-tracking system in Figure 6.2 becomes

$$L = \frac{GQ}{G_n(1 - Q)}. \tag{6.9}$$

Figure 6.2 DOB curvature rejection filter loop for path-tracking of an autonomous vehicle.

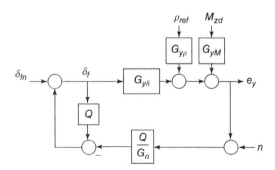

The new relation between the steering wheel input and the path-tracking error output becomes

$$\frac{e_y}{\delta_f} = \frac{L/(Q/G_n)}{1+L} = \frac{G_n G}{G_n(1-Q) + GQ}. \tag{6.10}$$

Desired path curvature rejection is given by

$$\frac{e_y}{\rho_{ref}} = \frac{G_{yp}}{1+L} = \frac{G_n(1-Q)G_{yp}}{G_n(1-Q) + GQ}. \tag{6.11}$$

Yaw moment disturbance rejection is given by

$$\frac{e_y}{M_{zd}} = \frac{G_{yM}}{1+L} = \frac{G_n(1-Q)G_{yM}}{G_n(1-Q) + GQ}. \tag{6.12}$$

Sensor noise rejection is given by

$$\frac{e_y}{n} = \frac{-L}{1+L} = \frac{GQ}{G_n(1-Q) + GQ}. \tag{6.13}$$

It is obvious from Eqs. (6.11) and (6.12) that if Q is designed as a unity gain low-pass filter, then desired path curvature rejection, i.e. tracking of the desired path, and yaw moment disturbance rejection will be achieved simultaneously within its bandwidth. Equation (6.13) merely shows that the path-following error value has the sensor noise in it at low frequencies where Q is close to unity. This shows that the DOB application is an effective inner loop that helps in following a desired path while also providing inherent yaw stability. At higher frequencies where Q approaches zero, sensor noise is not a consideration as DOB is effectively canceled and path-following error e_y is not being fed back. Q filter design is discussed in more detail in Section 6.3.

6.2.2 Application Examples

Four different implementations of the DOB structure for different applications in automated driving are discussed here. Each subsection presents a case study of a real-world problem along with the loop structure used. Simulation results are also shared, separately or comparatively, depending on the application. The vehicle model used here in these examples was described in detail in Chapter 2 and is, therefore, not shown again, but the transfer functions G and G_n of the actual and desired plant dynamics, respectively, are provided along with the vehicle parameters used.

Yaw Moment Disturbance Rejection During Path-Tracking

While the vehicle is driving autonomously, external effects such as wind or road conditions can cause the action of an unwanted yaw moment on the vehicle that is trying to follow a path. This effect can result in a small or large yaw motion which would be dangerous in traffic and should, therefore, be compensated. A yaw moment disturbance to sudden appearance of off-center side wind is illustrated in Figure 6.3. The current practice is to temporarily disengage path-tracking steering inputs and to temporarily give control authority to a yaw stability controller, also called electronic stability control, and to hand-control authority back to path-tracking control after the yaw response is stabilized. Stability is meant in the vehicle dynamics sense here, meaning not having excessive or oscillatory yaw rate response, and not in the control systems sense. Note that only the yaw moment disturbance caused by the off-center side wind is considered in this subsection while neglecting the sideways displacement that will be caused by the lateral wind force which would also be easily taken care of by the DOB loop.

Figure 6.3 Wind force causing yaw moment disturbance for the vehicle.

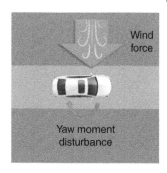

In Figure 6.2, if we set the steering wheel input to zero ($\delta_f = 0$) and the road curvature to zero ($\rho_{ref} = 0$), then the yaw moment disturbance M_{zd} will be stabilized by the vehicle using the DOB loop and path-tracking error e_y will go to zero after an initial transient. This is, in contrast, to the DOB applied to yaw stabilization alone [6], where the yaw rate is stabilized but the path-tracking error e_y does not go to zero. In Figure 6.2, the yaw moment disturbance rejection e_y/M_{zd} is given by Eq. (6.12). A Simulink model was constructed similar to Figure 6.2 to test the DOB path-tracking performance in the presence of the yaw moment disturbance. Vehicle parameters for the Ford Fusion Hybrid experimental vehicle discussed in Chapter 3 were used in the models for $G_{y\delta}$, G_n, and G_{yM} that were calculated using Eq. (2.95) and a state-space to transfer function transformation. This transformation was done for the relation between steering angle command δ_f and yaw rate r. It is important to note that our main goal in this subsection is testing the yaw moment disturbance rejection and not model regulation. Hence, we used identical transfer functions for $G_{y\delta}$ and G_n in the simulations that are given by

$$\frac{e_y}{\delta_f} = G_{y\delta} = G_n = \frac{227.7s^2 + 5542s + 36\,310}{s^4 + 22.19s^3 + 37.8s^2}. \tag{6.14}$$

The yaw moment disturbance rejection transfer function is given by

$$G_{yM} = \frac{0.000\,536\,5s^2 + 0.015\,35s + 0.092\,65}{s^4 + 22.19s^3 + 37.8s^2}. \tag{6.15}$$

Q was designed as a third-order filter and named Q_1. It is given by

$$Q_1 = \frac{0.45s + 1}{0.003\,375s^3 + 0.0675s^2 + 0.45s + 1}. \tag{6.16}$$

The design structure for the Q filters is discussed in Section 6.3. Its d.c. gain at $s = 0$ is unity as desired. Q_1 has a bandwidth of about 6 rad/s (close to 1 Hz) within which its gain is very close to 1 or 0 dB as seen in Figure 6.4. Note that we want the Q filter itself to be unity and not just its magnitude, so its phase should also be close to 0°. From Figure 6.4, it is seen that the phase of Q_1 in Eq. (6.16) is close to 0° up to a frequency of 1 rad/s. A higher bandwidth Q filter named Q_2 was also designed and is given by

$$Q_2 = \frac{0.15s + 1}{0.000\,125s^3 + 0.0075s^2 + 0.15s + 1}. \tag{6.17}$$

The magnitude and phase of filter Q_2 are also displayed in Figure 6.4 for comparison purposes. The disturbance rejection and model regulation properties of Q_2 are expected to be better that those of Q_1 at the expense of higher sensitivity to modeling error and high-frequency noise. These effects will get more pronounced as the bandwidth of the Q filter is increased.

Yaw moment M_{zd} pulses of 2500 N m magnitude were applied to the vehicle driving on a straight road ($\delta_f = 0$, $\rho_{ref} = 0$) at two seconds and at five seconds in the simulations, and the results are

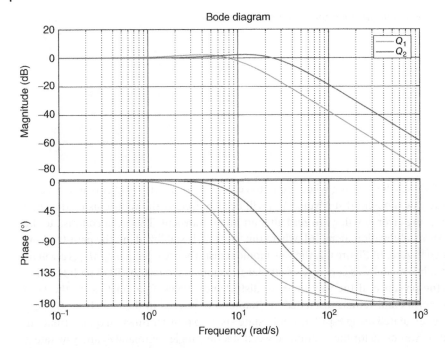

Figure 6.4 *Q* filter frequency responses.

shown in Figure 6.5. Filter Q_1 in Eq. (6.16) was used in the simulation. The simulation ran for 10 seconds while the vehicle was trying to move straight at a speed of 60 km/h. The yaw moment disturbance pulses can be due to sudden side wind and have been applied as disturbances with short spans of 0.25 seconds duration in both directions. Simulation results can be seen in Figure 6.5 and show that the DOB applies corrective steering action at the beginning and at the end of the yaw moment disturbance pulse to reduce the path-tracking error. The path-tracking error does not go to exactly zero after the transient due to the finite and limited bandwidth of the *Q* filter as the DOB compensation only works when *Q* is exactly unity. The pulse disturbance has signal content at high frequencies also which cannot be completely rejected as they are outside the *Q* filter bandwidth. The higher bandwidth Q_2 filter decreases this error as shown in the superimposed plots in Figure 6.5 at the expense of higher steering wheel actuation values.

Road Curvature Rejection

Desired path curvature rejection given by Eq. (6.11) can be used as an automatic path following system for autonomous vehicles. Figure 6.2 set the external steering wheel to 0 as $\delta_{fn} = 0$ and also set the yaw moment disturbance to 0 as $M_{zd} = 0$. Only the curvature $\rho_{ref} \neq 0$ will change in the simulations to show that the DOB will be able to reject curvature change as a disturbance, effectively following the desired curved path. Simulation results for several different desired paths are presented next to demonstrate how the DOB road curvature rejection works. It is important to note that, similar to the previous case study, since our only goal is testing the road curvature rejection and not the model regulation, identical transfer functions for $G_{y\delta}$ and G_n that were given in Eq. (6.14) were used in the simulation. The vehicle was set to move at the constant speed of 60 km/h. The *Q* filter was chosen as Q_1 given in Eq. (6.16).

The first desired path is a large oval shaped one generated using two parallel lines and two half circles. This creates a step-like change in curvature ρ_{ref} jumping from 0 at the straight part of the

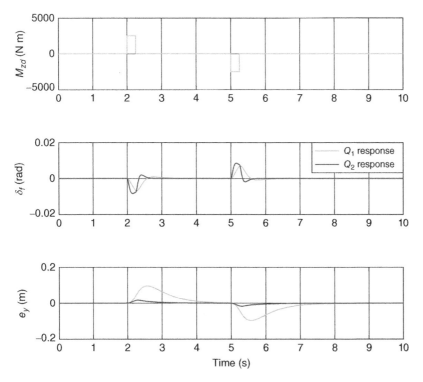

Figure 6.5 Simulation results for yaw moment disturbance rejection case study.

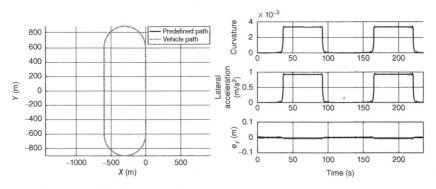

Figure 6.6 Simulation results for road curvature rejection case study, first path.

path to a constant value of 1 over the radius for the circular parts of the path. This is illustrated in the simulation results of Figure 6.6 which show the desired path, the corresponding curvature disturbance and the path following error. It is seen that DOB curvature rejection reduces the path-following error to very small values.

After the simulation, results were plotted as the vehicle path traveled and the path following error. These can be seen in Figure 6.6 as the vehicle was able to successfully follow the path with very small error. The error peaks occur while the vehicle is following the curved parts of the path.

The second desired path is a racetrack with a more significant amount of curvature variation. The vehicle path, the corresponding path curvature disturbance and the path following error are

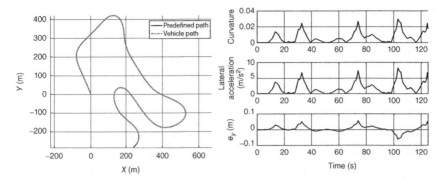

Figure 6.7 Simulation results for road curvature rejection case study, second path.

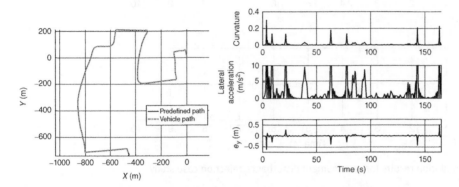

Figure 6.8 Simulation results for road curvature rejection case study, third path.

shown in Figure 6.7. Looking at the simulation results, the vehicle can be seen to be successfully following the path with a small amount of error but larger as compared to the simpler oval path in Figure 6.6 with error being large at locations of high curvature again.

The third desired path is a more realistic one taken from a real-world university campus environment. The desired path, the corresponding path curvature disturbance, and the path-tracking error are shown in Figure 6.8. Note that the chosen 60 km/h constant speed is higher than the maximum allowed speed for this path. Even then, the DOB performs well at rejecting the curvature to follow the path correctly and with acceptably small tracking error overall. However, lateral acceleration is occasionally very high for a short time around the sharp turns. As expected, path-tracking error is also highest around sharp turns. In a real-world scenario, a speed controller would be present to reduce the speed while approaching the sharper turns. The path-tracking error can be reduced by decreasing speed while entering and taking curves and by using a higher bandwidth Q filter. It should be noted, however, that the Q filter bandwidth is limited by the bandwidth of the steering actuation system. As a result of this case study, we can see that using DOB in the loop shown in Figure 6.2, with the model described, produces good results for path-tracking. The vehicle is seen to be able to handle all of the desired paths that have increasingly more difficult road-curvature disturbance inputs at relatively high speed.

Road Curvature Rejection with DOB within a Control Loop

In this case study, we will look into the same problem as in the Road Curvature Rejection application example. In contrast to the Road Curvature Rejection application example, however,

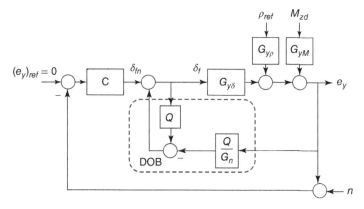

Figure 6.9 Loop structure for road curvature rejection with DOB in a control loop.

we augment the DOB with a feedback controller to reduce any remaining path-tracking errors. This path-tracking control augmented version of the DOB compensated path-tracking model is shown in Figure 6.9. The extra feedback loop and the feedback controller are expected to reduce the path-tracking error more as compared to the DOB compensation which will have Q filter bandwidth limitations due to stability robustness in the presence of model uncertainty and in order not to respond to sensor noise.

Block diagram algebra in Figure 6.9 results in its equivalent form in Figure 6.10 using which the following transfer functions for this closed loop case can easily be obtained.

Since the desired path following error is equal to zero, command or reference following does not make sense and is skipped. The relation between the path-curvature disturbance and the path-tracking error output becomes

$$\left(\frac{e_y}{\rho_{ref}}\right)_{cl} = \frac{G_n G_{y\rho}(1-Q)}{G_n(1-Q) + G_{y\delta}(CG_n + Q)}, \tag{6.18}$$

where the subscript cl on the left-hand side shows that this is for the closed-loop system in Figure 6.9. The yaw moment disturbance rejection is given by

$$\left(\frac{e_y}{M_{zd}}\right)_{cl} = \frac{G_n G_{yM}(1-Q)}{G_n(1-Q) + G_{y\delta}(CG_n + Q)}. \tag{6.19}$$

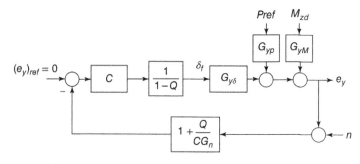

Figure 6.10 Equivalent form of DOB compensated path-tracking loop under feedback control.

Sensor noise rejection is given by

$$\left(\frac{e_y}{n}\right)_{cl} = \frac{-G_{y\delta}(CG_n + Q)}{G_n(1 - Q) + G_{y\delta}(CG_n + Q)}. \tag{6.20}$$

The ideal case of $Q = 1$ at low frequencies results in perfect path curvature rejection in Eq. (6.18) and perfect yaw moment disturbance rejection in Eq. (6.19). For $Q = 1$ at low frequencies, the sensor noise is reflected directly in the path-tracking error. $Q = 0$ at high frequencies means that the DOB is cut off and the sensor noise is reflected just like the case for the plant under feedback control without the DOB, i.e. it is reduced due to the sensor noise rejection properties of the feedback control loop alone.

The simulations for the three different desired paths in the Road Curvature Rejection application example are repeated here for the closed-loop controlled DOB loop. For the simulations, same conditions, predefined paths, and transfer functions as the previous case study were used for ease of comparison. The feedback steering controller was hand tuned to obtain

$$C(s) = \frac{0.35s^2 + 0.15s}{s} \tag{6.21}$$

and is a Proportional-Derivative (PD) controller. Simulation results are shown below, starting from the first oval path in Figure 6.11. Desired curvature is not shown as it was shown in the corresponding plots in the Road Curvature Rejection application example. Looking at the error graph, there is almost no visible tracking error. Compared to the previous case study with results shown in Figure 6.6, this is a good improvement that is obtained by just adding a hand-tuned steering controller into the loop.

For the second racetrack-like path with results shown in Figure 6.12, although slightly more visible compared to the first path, the error is still very small. Compared to the previous case study results in Figure 6.7, there is a significant improvement overall.

For the third real university campus route path with results show in Figure 6.13, again, results are improved significantly as compared to the previous case study results for the same path that were shown in Figure 6.8. Results are improved both in terms of reduced error magnitude and in terms of faster response, as the error dies down very quickly.

Looking at the overall results and comparing with the previous case study, where DOB was used by itself as the curvature rejection filter, using DOB within a feedback control loop provides much better results. Moreover, the controller does not have to be well tuned for the system to improve the performance which makes this method much easier to implement as compared to other methods [14, 15].

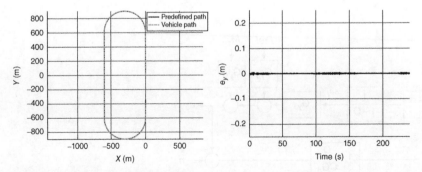

Figure 6.11 Simulation results for road curvature rejection with DOB in a control loop case study, first path.

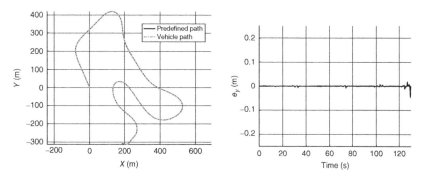

Figure 6.12 Simulation results for road curvature rejection with DOB in a control loop case study, second path.

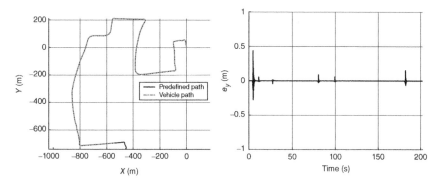

Figure 6.13 Simulation results for road curvature rejection with DOB in a control loop case study, third path.

Model Regulation with DOB

While designing the control structure for the autonomous vehicle, vehicle parameters play an important role. When implemented on the real vehicle after the design and simulations, correctly tuned controllers might not produce the desired results or can even make the vehicle unstable if there is a large mismatch between modeled and actual plant dynamics. In these types of situations, or if we want to regulate the plant to a desired model, the DOB can be used to reject the parameter changes in the system $G_{y\delta}$ as compared to its nominal or desired model G_n. It should be noted that there is a stability robustness based limit on the difference between the actual and desired or nominal plants that can be accommodated using the DOB [5, 6, 16]. We will analyze this feature in more detail in this case study. The DOB loop with no external disturbances and noise is repeated below in Figure 6.14. The path curvature and yaw moment disturbances are not shown in Figure 6.14

Figure 6.14 Loop structure for model regulation with DOB case study.

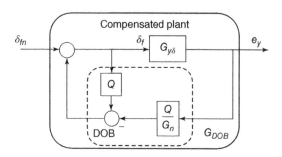

since we are only interested in investigating the model regulation properties of the DOB loop in this subsection.

Considering the DOB compensated plant in Figure 6.14, it can be seen that the DOB compensated plant G_{DOB} becomes

$$G_{DOB} = \frac{e_y}{\delta_{fn}} = \frac{G_n G_{y\delta}}{G_n(1 - Q) + G_{y\delta}Q}, \tag{6.22}$$

which becomes the nominal or desired plant G_n when Q is equal to unity. Using this relationship, we can investigate how well this model regulation works for our nominal or desired plant G_n, compensated plant G_{DOB}, and the actual plant $G_{y\delta}$ with mass and tire coefficients modified. These three systems were analyzed in terms of frequency response using Bode plots as shown in Figure 6.15. The Q filter used is also shown so that the reader can take a look at the frequency range where Q has unity gain and close to zero-phase angle. The nominal model G_n in Eq. (6.14) was used. The actual plant $G_{y\delta}$ was modified by having 1.5 times more mass and tire cornering stiffness coefficients and is given by

$$G_{y\delta} = \frac{294s^2 + 8314s + 54\,460}{s^4 + 22.92s^3 + 56.7s^2}. \tag{6.23}$$

As it can be seen from the frequency response, looking at the similarities between the dashed line for G_{DOB} and line for G_n, from the low-to-mid frequency, the DOB is seen to successfully make the actual plant behave like its nominal model for up to 20 rad/s for magnitude and 5 rad/s for phase. Within the mentioned frequency ranges, the magnitude and phase frequency responses for the compensated plant are very similar to those of the nominal plant. This compensation is not possible at high frequencies, but those high frequencies are usually not within the operating region

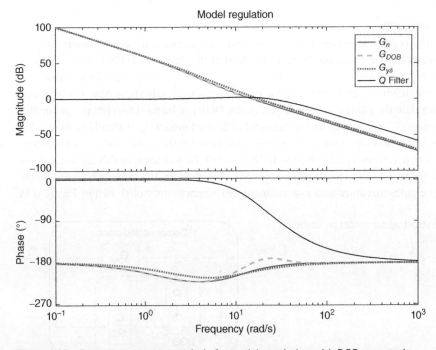

Figure 6.15 Frequency response analysis for model regulation with DOB case study.

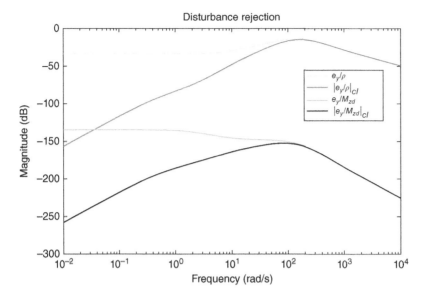

Figure 6.16 Disturbance rejection properties comparison for disturbances within and out of a control loop.

for the purpose of path-tracking. If necessary, the bandwidth of the *Q* filter can be increased to improve the frequency range of model regulation.

6.2.3 Disturbance Rejection Comparison

Using the equations derived in the case studies, disturbance rejection properties of the designed DOB loops were plotted with respect to frequency, on top of each other for comparison. These plots show road curvature rejection $|e_y/\rho|$ and $|e_y/\rho|_{cl}$ without and with a feedback controller for path-tracking. They also show yaw moment disturbance rejection $|e_y/M_{zd}|$ and $|e_y/M_{zd}|_{cl}$ without and with a feedback controller for path-tracking. These plots displayed together in Figure 6.16 show good road curvature and yaw moment disturbance rejection at low frequencies. If direct comparison could be made, yaw moment disturbance rejection would be seen to be several orders of magnitude better than road curvature disturbance rejection. Closed-loop feedback control also improves disturbance rejection considerably in both cases.

6.3 *Q* Filter Design

For specific choices of $Q(s)$, the model regulator introduces integrators into the loop, resulting in reduced steady-state error [16]. Example first-, second-, and third-order generic *Q* filters and the number of free integrators they introduce to the DOB loop are given below as

$$Q_{fo}(s) = \frac{1}{(\tau s + 1)^l}, \quad \text{(1 free integrator)}, \tag{6.24}$$

$$Q_{so}(s) = \frac{1}{\left(\frac{s}{\omega}\right)^2 + 2\zeta\frac{s}{\omega} + 1}, \quad \text{(1 free integrator)}, \tag{6.25}$$

$$Q_{to}(s) = \frac{3\tau s + 1}{\tau^3 s^3 + 3\tau^2 s^2 + 3\tau s + 1}, \quad \text{(2 free integrators)}, \tag{6.26}$$

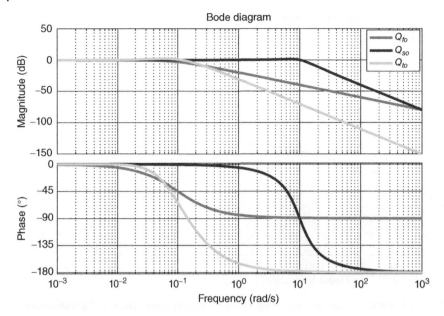

Figure 6.17 Different Q filter frequency responses.

where ω or τ can be determined depending on desired filter properties. Figure 6.17 shows frequency responses for the different Q filters in Eqs. (6.24)–(6.26) with different cutoff frequencies for being able to distinguish between them on the same plot, using values of $\tau = 10, \zeta = 5, \omega = 100$.

6.4 Time Delay Performance

The path-tracking control system uses the steering actuator which is controlled by issuing CAN commands in current drive-by-wire vehicles. This CAN bus communication introduces time delay into the path-tracking model which will degrade performance. It is known that large amounts of time delay in a system with feedback will destabilize the system due to the extra-negative phase injected into the loop due to the presence of the time delay. There are well-known time delay compensation schemes like the Smith Predictor if the time delay value is exactly known [17]. Similarly, the CDOB can be used when the time delay value is unknown [16]. Time delay cannot be canceled by either of these methods that aim to effectively move the time delay outside the feedback loop. The response is still delayed but the feedback controller is not affected by it and is designed for the corresponding model without time delay, and the system is stable regardless of the time delay. The effect of the delay can be reduced by looking at the future of the reference input which is similar to reference feedforward control [18]. None of these methods of the Smith Predictor or CDOB can be directly applied to road vehicle path-tracking control as it is formulated as a disturbance rejection problem and not one of reference input tracking. It is assumed that the CAN bus-based delay in the steering actuation system are bounded and well characterized and is treated as an uncertainty. It turns out that the DOB loop inherently has excellent time delay compensation properties. The rest of this section investigates the effect of this model uncertainty on the performance of the DOB-based curvature rejection filter used by itself and also if used within a feedback control loop as presented in Section 6.3.

Figure 6.18 DOB curvature rejection filter with steering actuator time delay.

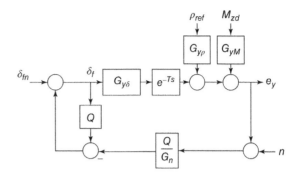

Figure 6.18 shows the DOB curvature rejection filter of Figure 6.2. The block e^{-Ts} in Figure 6.18 models the steering actuator delay of T seconds. Note that road curvature changes and yaw moment disturbances affect the path following error directly. Only the steering corrections that are needed to keep the vehicle on the desired path go through the actuator dynamics and the associated time delay.

The DOB loop transfer functions in Figure 6.18 are investigated next. Model regulation becomes

$$\frac{e_y}{\delta_f} = \frac{G_n G e^{-Ts}}{G_n(1-Q) + QGe^{-Ts}}. \tag{6.27}$$

In the case of perfect model regulation for $Q = 1$ at low frequencies, this equation becomes

$$\frac{e_y}{\delta_f} = G_n \tag{6.28}$$

as expected and desired. Note that Eq. (6.28) is independent of the time delay as long as $Q = 1$. As $Q \to 0$ at high frequencies, the model goes back to Ge^{-Ts} as it should. The desired path curvature rejection is given by

$$\frac{e_y}{\rho_{ref}} = \frac{G_n(1-Q)G_{y\rho}}{G_n(1-Q) + QGe^{-Ts}}, \tag{6.29}$$

which still goes to zero as

$$\left.\frac{e_y}{\rho_{ref}}\right|_{Q=1} = 0 \tag{6.30}$$

for $Q = 1$ at low frequencies. As $Q \to 0$ at high frequencies, the path curvature rejection becomes $e_y/\rho_{ref} \to G_{y\rho}$ as it should be. Yaw moment disturbance rejection becomes

$$\frac{e_y}{M_{zd}} = \frac{G_n(1-Q)G_{yM}}{G_n(1-Q) + QGe^{-Ts}} \tag{6.31}$$

which becomes zero as

$$\left.\frac{e_y}{M_{zd}}\right|_{Q=1} = 0 \tag{6.32}$$

for $Q = 1$ at low frequencies. As $Q \to 0$ at high frequencies, the path curvature rejection becomes $e_y/M_{zd} \to G_{yp}$ as it should be. There is, of course, a limit on this excellent time delay compensation due to stability robustness as the time delay is treated as a model uncertainty.

Consider steering model uncertainty as

$$G_{y\delta} = G_n(1 + \Delta_m)e^{-Ts}, \tag{6.33}$$

where Δ_m is multiplicative uncertainty that does not include the time delay. The total multiplicative uncertainty is calculated as

$$\frac{G_{y\delta} - G_n}{G_n} = \frac{G_n(1 + \Delta_m)e^{-Ts} - G_n}{G_n} = (1 + \Delta_m)e^{-Ts} - 1. \tag{6.34}$$

A conservative condition for stability robustness of the DOB curvature rejection filter loop in Figure 6.18, then, becomes (see [16])

$$|Q| < \frac{1}{|(1 + \Delta_m)e^{-Ts} - 1|}. \tag{6.35}$$

A relatively large time delay value of 80 ms for the steering actuation is used to illustrate the time delay performance of the DOB curvature rejection filter. This relatively large delay is typical of a retrofit drive-by-wire system which is currently used in most research autonomous vehicles. Condition (6.35) is illustrated for this time delay with no other model uncertainty, i.e. $\Delta_m = 0$, in Figure 6.19. Magnitude plots for two different filters Q_1 and Q_2 are shown along with the right-hand side of (6.35). As seen in the plot, due to its higher cut-off frequency, Q_2 intersects with the frequencies where the right-hand side of (6.35) is getting small. Condition (6.35) is satisfied for filter Q_1. Simulations conducted with the DOB loop using Q_1 and Q_2 show that while the system gets unstable if Q_2 is used, the path following can still be performed with the delay present if Q_1 is used, since it meets the stability criteria. Figure 6.20 shows the simulation results for the third path considered in earlier simulations with Q_1 and a delayed system. Comparison of results to those in Figure 6.8 show that good path-tracking is achieved even though the results are, as expected, not as good as the previous ones due to the presence of the time delay and stability robustness limitations on the bandwidth of the Q filter that can be used. Figure 6.21 shows the corresponding simulation result where a feedback loop has also been added with the same feedback controller that was used before. The results are good again, although there is some deterioration due to the presence of the steering actuation time delay.

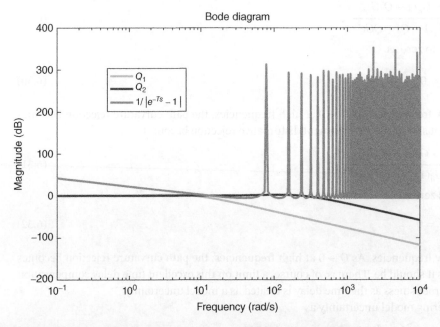

Figure 6.19 Illustration of stability robustness of DOB against time delay.

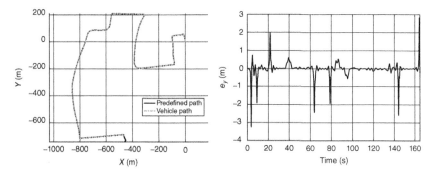

Figure 6.20 Simulation results for road curvature rejection with DOB with time delay, third path.

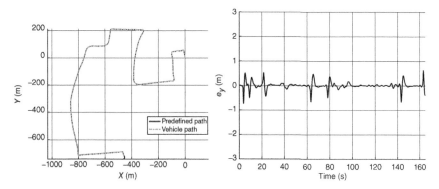

Figure 6.21 Simulation results for road curvature rejection with DOB with time delay and feedback control, third path.

6.5 Chapter Summary and Concluding Remarks

In this chapter, model regulation also called the disturbance observer (DOB) was introduced as a path curvature rejection filter for autonomous vehicle path following. Several application examples were presented to demonstrate the different benefits of DOB in several cases applicable to automated driving. This was followed by disturbance rejection comparison for different configurations and Q filter design information. Time delay compensation properties of the DOB were discussed last along with examples that demonstrated successful compensation.

References

1 B. A. Guvenc, L. Guvenc, E. S. Ozturk and T. Yigit (2003) Model regulator based individual wheel braking control, *Proceedings of 2003 IEEE Conference on Control Applications, 2003. CCA 2003*, vol. 1, pp. 31–36. https://doi.org/10.1109/CCA.2003.1223254.

2 H. Wang and L. Guvenc, "Use of robust DOB/CDOB compensation to improve autonomous vehicle path following performance in the presence of model uncertainty, CAN bus delays and external disturbances," *WCX18: SAE World Congress Experience*, IEEE, 2018, 31–36.

3 K. Ohnishi, "A new servo method in mechatronics," *Transactions of Japanese Society of Electrical Engineering, D*, vol. 107-D, pp. 83–86, 1987.

4 T. Umeno and Y. Hori, "Robust speed control of DC servomotors using modern two degrees-of-freedom controller design," *IEEE Transactions on Industrial Electronics*, vol. 38, no. 5, pp. 363–368, 1991.

5 L. Güvenç and K. Srinivasan, "Friction compensation and evaluation for a force control application," *Mechanical Systems and Signal Processing*, vol. 8, no. 6, pp. 623–638, 1994.

6 B. A. Guvenc, L. Guvenc, and S. Karaman, "Robust yaw stability controller design and hardware-in-the-loop testing for a road vehicle," *IEEE Transactions on Vehicular Technology*, vol. 58, no. 2, pp. 555–571, 2009.

7 B. A. Güvenç, L. Güvenç, and S. Karaman, "Robust MIMO disturbance observer analysis and design with application to active car steering," *International Journal of Robust and Nonlinear Control: IFAC-Affiliated Journal*, vol. 20, no. 8, pp. 873–891, 2010.

8 B. Aksun Güvenç, S. Necipoğlu, B. Demirel, L. Güvenç, 2011, "Chapter 3: Robust control of atomic force microscopy," *Mechatronics*: P. Davim, ISTE/Wiley, London, ISBN: 978-1-84821-308-1, pp. 103–132.

9 X. Fan and M. Tomizuka, "Robust disturbance observer design for a power-assist electric bicycle," in *Proceedings of the 2010 American Control Conference*, 2010, pp. 1166–1171: IEEE.

10 T. Satoh, K. Kaneko, and N. Saito, "Improving tracking performance of predictive functional control using disturbance observer and its application to table drive systems," *International Journal of Computers Communications & Control*, vol. 7, no. 3, pp. 550–564, 2014.

11 Y. Luo, T. Zhang, B. Lee, C. Kang, and Y. Chen, "Disturbance observer design with Bode's ideal cut-off filter in hard-disc-drive servo system," *Mechatronics*, vol. 23, no. 7, pp. 856–862, 2013.

12 K. Natori and K. Ohnishi, "A design method of communication disturbance observer for time-delay compensation, taking the dynamic property of network disturbance into account," *IEEE Transactions on Industrial Electronics*, vol. 55, no. 5, pp. 2152–2168, 2008.

13 W. Zhang; M. Tomizuka; P. Wu; Y.-H. Wei; Q. Leng; S. Han; A.K. Mok *et al.*, "A double disturbance observer design for compensation of unknown time delay in a wireless motion control system," *IEEE Transactions on Control Systems Technology*, vol. 26, no. 2, pp. 675–683, 2018, https://doi.org/10.1109/TCST.2017.2665967.

14 M. R. Cantas and L. Guvenc, "Camera based automated lane keeping application complemented by GPS localization based path following," SAE World Congress and Experience, 2018.

15 O. Tuncer, L. Guvenc, F. Coskun, and E. Karsligil, "Vision based lane keeping assistance control triggered by a driver inattention monitor," *IEEE International Conference on Systems, Man and Cybernetics*, IEEE, 2010, 289–297.

16 L. Guvenc, B. Aksun-Guvenc, B. Demirel, and M. T. Emirler, "*Control of Mechatronic Systems*," IET, London, 2017.

17 O. J. M. Smith, "A controller to overcome dead time," *ISA Journal*, vol. 6(2), pp. 28–33, 1959.

18 H. Wang, A. Tota, B. Aksun-Guvenc, and L. Guvenc, "Real time implementation of socially acceptable collision avoidance of a low speed autonomous shuttle using the elastic band method," *Mechatronics*, vol. 50, pp. 341–355, 2018.

7

Robust Path Tracking Control

7.1 Introduction

This chapter starts with the model predictive control (MPC) method due to its widespread use in automotive control applications and increasing use in path-tracking control. MPC predicts the vehicle state according to the inherent model and compares future output with future reference to determine the control action while simultaneously penalizing a chosen length of future control action. This preview feature makes MPC steering control smooth and thereby relaxes the requirements on the path description. Its performance in path following control is evaluated through both simulations and vehicle experiments in this chapter.

During autonomous driving, operating conditions such as the vehicle load and traveling speed change. Tire characteristics will also vary substantially under different weather and road conditions. This limits the maximum values of the longitudinal and lateral forces that the tires can provide. Based on these considerations, a first requirement on the control law for path following is that the controlled system must remain stable for the different possible operating conditions. This is a robustness of stability requirement and includes nominal stability in the case of no uncertainty. Further guarantee of robustness of performance in the presence of defined uncertainty regions is also highly desirable. This is a robustness of performance requirement which includes nominal performance in the absence of uncertainty. We want to determine or tune control parameters analytically to achieve robust stability and performance and present a robust gain-scheduling controller design based on the parameter-space approach for this purpose. The gain-scheduling control is of interest for systems with varying dynamics at different operating conditions. Requirements of robustness of stability and performance are also necessary if the system is subject to other parametric or model uncertainties. The parameter space approach is capable of projecting these requirements to the space of design parameters, which makes the design of robust gain-scheduling control more intuitive. Multiobjective controller design is also very easy with this approach. The fundamental limitation in parameter space control methods is being able to concentrate on only two uncertain parameters or control gains (or a combination of one of each). This method is applied to automated vehicle lateral control to ensure the tracking accuracy and stability subject to uncertainties in vehicle load, speed, and tire saturation. The proportional feedback gain and look-ahead distance are scheduled with respect to speed while sustaining robust stability, mixed sensitivity bound constraint, and satisfaction of performance indices. Realistic hardware-in-the-loop (HIL) simulation with a validated vehicle model and road tests are conducted to demonstrate the robust performance of the designed gain-scheduling control in the presence of model uncertainty and disturbances. Another general robust gain-scheduling approach, the linear-matrix-inequality (LMI) design is also used later for benchmarking purposes and the comparisons are discussed.

Autonomous Road Vehicle Path Planning and Tracking Control, First Edition.
Levent Güvenç, Bilin Aksun-Güvenç, Sheng Zhu, and Şükrü Yaren Gelbal.
© 2022 The Institute of Electrical and Electronics Engineers, Inc. Published 2022 by John Wiley & Sons, Inc.

Moreover, we also consider the case when the planned trajectory is not dynamically feasible for the path following control. The improper trajectory planning may not be responsive in a short time due to estimation error of the road friction coefficient when the vehicle just enters a slippery road section and the estimation module does not have enough time or enough longitudinal excitations. To reduce safety risk in that case, it is necessary to consider adding the vehicle stability control (VSC) to further enhance driving safety for autonomous driving. The main idea of the VSC is to compensate for undesired yaw motions by providing corrective yaw moment in order to maintain the vehicle yaw stability. VSC is not a new technology, but we are interested in its integration with the autonomous driving functions and to observe its potential benefits in both yaw stability and path-tracking performance. We will discuss two strategies in deciding the necessary amount of direct yaw moment compensation and their implementations through differential braking.

7.2 Model Predictive Control for Path Following

7.2.1 Formulation of Linear Adaptive MPC Problem

The MPC is first investigated due to its gaining popularity in the application of path following control along with increasing computational power of on-board computers making its direct, online implementation possible. In the path following control applications, methods like robust proportional integral derivative (PID) [1], and robust gain-scheduling [2–4] determine the steering angle based on feedback of current tracking errors, while simple geometric/kinematic pure-pursuit based techniques [5, 6] merely target on a specific future waypoint along the planned path. Consequently, these approaches cannot take full advantage of the future path information. MPC, on the other hand, differs from the above methods by considering a continuous period of future reference path values to determine the steering control action, which generally delivers better tracking accuracy with less-steering oscillations. Due to this feature, requirements on the reference path planning are also less stringent for the MPC controller as compared to other control methods. Unlike other path descriptions such as splines [7], to achieve continuous change of heading and position, a simple-path representation in sparse discrete waypoints is enough for the MPC controller as shown in this section. These features make MPC a suitable choice to track the discrete waypoints modified by online planning techniques such as the elastic-band method for collision avoidance purposes. One challenge in implementing MPC is its relatively large computational complexity due to the optimization involved at each computation instance, thus, making its real-time applications uncommon especially in time-sensitive occasions. In the following work presented in this section, an adaptive MPC control strategy based on a linearized dynamic model is used to reduce its computational burden.

The model for the MPC lateral tracking was formulated in a moving vehicle framework as shown in Figure 7.1. For a short preview horizon, it is acceptable to assume that the vehicle travels at a constant speed v_x. Under this assumption, the vehicle lateral dynamics can be described by the following state-space representation:

$$\dot{\xi} = A_c\xi + B_cu, \tag{7.1}$$
$$\eta = C_c\xi, \tag{7.2}$$

Figure 7.1 Preview references in the moving-vehicle framework. Source: Zhu and Aksun-Guvenc [8]/with permission from Springer Nature.

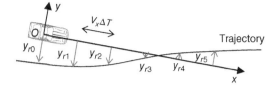

where the state-space description matrices are

$$A_c = \begin{bmatrix} 0 & 1 & 0 & 0 \\ 0 & -(C_{af} + C_{ar})/mv_x & (C_{af} + C_{ar})/m & (l_r C_{ar} - l_f C_{af})/mv_x \\ 0 & 0 & 0 & 1 \\ 0 & (l_r C_{ar} - l_f C_{af})/I_z v_x & (l_f C_{af} - l_r C_{ar})/I_z & -\left(l_f^2 C_{af} + l_r^2 C_{ar}\right)/I_z v_x \end{bmatrix}, \tag{7.3}$$

$$B_c = \begin{bmatrix} 0 \\ C_{af}/m \\ 0 \\ l_f C_{af}/I_z \end{bmatrix}, \tag{7.4}$$

and

$$C_c = \begin{bmatrix} 1 & 0 & 0 & 0 \\ 0 & 0 & 1 & 0 \end{bmatrix}. \tag{7.5}$$

The control u is the front-wheel steering angle. The states $\xi = [y, \dot{y}, \psi, \dot{\psi}]$ are chosen as the vehicle's lateral coordinate y, heading ψ, and their derivatives. Equation (7.1) was derived from the bicycle model as in [9, 10] and also presented in detail in Chapter 2, with C_{af} and C_{ar} denoting the lumped tire cornering stiffness values at the front and rear axles, respectively. Similarly, l_f and l_r the lengths from the vehicle center of mass to its front and rear axle, respectively. m is the vehicle mass, and I_z is the vehicle moment of inertia about the vertical z axis at its center of mass.

The parameter values of the state matrices are varying with the change of vehicle speed. Hence, the continuous state-space representation (7.1) needs to be updated at each time step and then converted to the discrete form at sampling time ($\Delta T = 0.04$ seconds here) as

$$\xi_{k+1} = A_d \xi_k + B_d[u_{k-1} + \Delta u_k], \tag{7.6}$$

$$\eta_k = C_d \xi_k. \tag{7.7}$$

Reference inputs $\eta_{ref} = [y_r; \psi_r]^T$ with y_r and ψ_r represent the lateral distance and vehicle orientation errors with respect to the current vehicle x axis of the path in the preview horizon, respectively. As illustrated in Figure 7.1, y_r are obtained at evenly distributed abscissa under the constant velocity assumption. The orientation errors ψ_r similarly are the errors of the path tangent from the current vehicle heading at these look-ahead positions. The cost function was defined to penalize deviation from the reference path and control effort Δu_k over the preview horizon from time step k:

$$J(\xi_k, \Delta \mathbf{u}) = \sum_{i=1}^{h_p} \|\hat{\eta}_{k+i|k} - \eta_{ref\ k+i|k}\|_Q^2 + \sum_{i=0}^{h_c-1} \|\Delta u_{k+j|k}\|_R^2, \tag{7.8}$$

where $\{\Delta u_{k|k}, \ldots, \Delta u_{k+h_c-1|k}\}$ is the control sequence to be optimized over the control horizon h_c and $\{\hat{\eta}_{k+1|k}, \hat{\eta}_{k+2|k}, \ldots, \hat{\eta}_{k+h_p|k}\}$ are the predicted states over the preview or prediction horizon h_p

after applying the control sequence. The control horizon h_c and the prediction horizon h_p are not necessarily the same. The control horizon h_c is usually much smaller than the prediction horizon h_p for a practical implementation in order to make the optimization of the cost function in Eq. (7.8) with respect to all the control values more tractable. Q and R are positive semidefinite and positive definite weighting matrices for reference tracking errors and steering control effort, respectively.

The MPC solves the finite horizon optimal control problem at every time step to minimize the cost in (7.8):

$$\min_{\Delta \mathbf{u}} J(\xi_k, \Delta \mathbf{u}), \tag{7.9}$$

$$\text{s.t. } \xi_{k+i+1|k} = A_d \xi_{k+i|k} + B_d [u_{k+i-1|k} + \Delta u_{k+i|k}], \tag{7.10}$$

$$\eta_{k+i+1|k} = C_d \xi_{k+i+1|k}, \tag{7.11}$$

$$\delta_{f,min} \leq u_{k+i|k} \leq \delta_{f,max}, \tag{7.12}$$

$$\Delta \delta_{f,min} \leq \Delta u_{k+i|k} \leq \Delta \delta_{f,max}, \quad i = 0,1, \ldots, h_p - 1, \tag{7.13}$$

where the control saturation and rate limits were also considered.

This properly formulated MPC problem with a linear adaptive model and a quadratic-form objective function can be solved efficiently using quadratic programming algorithms. It has been experimentally implemented in real-time at the frequency of 25 Hz in our work presented in this section. The first value $\Delta u^*_{k|k}$ in the optimal control sequence is used as the steering control for the current time step.

7.2.2 Estimation of Lateral Velocity

As stated before, the local ground-fixed coordinate Oxy is repositioned at each computation instance to align itself with the current vehicle position and posture as shown in Figure 7.1. This coordinate system is fixed during the prediction computations of the cost function in Eq. (7.8) and moves to the next position of the vehicle at the next computation instance. Therefore, the state values ξ_k under this framework become

$$\xi_k = [0, v_y, 0, r], \tag{7.14}$$

$\xi = [y, \dot{y}, \psi, \dot{\psi}]$ with $y = 0$ and $\psi = 0$ at the beginning of the kth step since the position error and pose error are both zero by construction, $\dot{y} = v_y$ which is the vehicle lateral speed and $\dot{\psi} = \omega_z = r$ which is the vehicle yaw rate. The vehicle yaw rate r can be obtained from the inertial measurement unit (IMU) sensor equipped in our test vehicle, but v_y cannot be directly measured. Therefore, a state estimation is necessary to make the state values ξ_k available for the successive MPC controller computations. Note that electronic stability control (ESC) is mandatory in all new vehicles and ESC has an inherent yaw rate sensor and a steering wheel angle sensor whose values can be used in control algorithms.

The Kalman filter has been commonly adopted in road vehicle estimation problems. It assumes the true state at time k has the following evolving relationship with itself from previous step in discrete form:

$$\mathbf{x}[k] = F_k \mathbf{x}[k-1] + B_k \mathbf{u}[k] + \mathbf{w}[k], \tag{7.15}$$

where F_k is the state-transition matrix applied on the previous state $\mathbf{x}[k-1]$, B_k is the input matrix applied to the control vector $\mathbf{u}[k]$, and $\mathbf{w}[k]$ is the process noise which is assumed to be distributed normally with covariance Q_k, $\mathbf{w}[k] \sim N(0, Q_k)$. The observation made at time k with respect to the true state is assumed to be

$$\mathbf{y}[k] = H_k \mathbf{x}[k] + \mathbf{v}[k], \tag{7.16}$$

where $\mathbf{v}[k]$ is the observation/measurement noise that follows the zero mean Gaussian distribution with covariance R_k, $\mathbf{v}[k] \sim N(0, R_k)$.

The Kalman filter could be conceptualized into two distinct phases: prediction and update. The prediction phase uses state estimates from the previous time step to estimate the state at the current time step and predict the state estimate covariance P:

$$\hat{\mathbf{x}}[k|k-1] = F_k \hat{\mathbf{x}}[k-1|k-1] + B_k \mathbf{u}[k], \tag{7.17}$$

$$\hat{P}[k|k-1] = F_k P[k-1|k-1] F_k^T + Q_k, \tag{7.18}$$

where the index expression "$[m|n]$" represents the estimate at time step m inferred from past measurements up to time step n. This predicted state estimate $\hat{\mathbf{x}}[k|k-1]$ is also named the priori state estimate since it does not include observation or measurement at current time step k.

The subsequent update phase further refines the state estimate by comparing the current measurement with the priori measurement estimate $\hat{\mathbf{y}}[k|k-1]$. The difference of the two is referred to as the innovation or measurement residual,

$$\tilde{\mathbf{y}}[k] = \mathbf{y}[k] - H_k \hat{\mathbf{x}}[k|k-1]. \tag{7.19}$$

The update state estimate $\hat{\mathbf{x}}[k|k]$, also called the posteriori state estimate, is then obtained by adding the correction term from the innovation:

$$\hat{\mathbf{x}}[k|k] = \hat{\mathbf{x}}[k|k-1] + K[k]\tilde{\mathbf{y}}[k], \tag{7.20}$$

where $K[k]$ is the Kalman gain. For the linear type Kalman filter, the Kalman gain $K[k]$ is optimal in the sense of minimizing the expected square of norm for the state estimate error $E(\|x[k] - \hat{x}[k|k]\|)$. The gain $K[k]$, along with the innovation covariance $S[k]$, the posterior state covariance, are updated iteratively as [11]:

$$S[k] = H_k P_{k|k-1} H_k^T + R_k, \tag{7.21}$$

$$K[k] = P_{k|k-1} H_k^T S^{-1}[k]. \tag{7.22}$$

During the lateral velocity estimation, the states are chosen as $\mathbf{x} = [v_y, r]'$, input $u = \delta_f$ and the measurement $y = r$. The state equation and measurement equation of the plant model used in the Kalman filter are

$$\dot{\mathbf{x}} = A_{kf}\mathbf{x} + B_{kf}u, \tag{7.23}$$

$$y = C_{kf}\mathbf{x}, \tag{7.24}$$

where

$$A_{kf} = \begin{bmatrix} -(C_{af} + C_{ar})/mv_x & (C_{ar}l_r - C_{af}l_f)/mv_x - v_x \\ (C_{ar}l_r - C_{af}l_f)/I_z v_x & -\left(C_{af}l_f^2 + C_{ar}l_r^2\right)/I_z v_x \end{bmatrix}, \tag{7.25}$$

$$B_{kf} = \begin{bmatrix} C_{af}/m \\ C_{af}l_f/I_z \end{bmatrix}, \quad C_{kf} = [0 \ 1]. \tag{7.26}$$

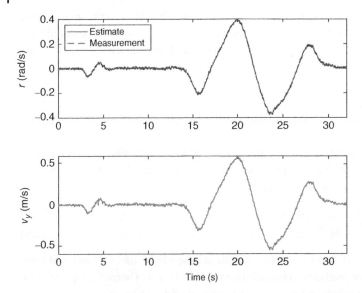

Figure 7.2 Data of yaw rate measurement and Kalman filter estimation results during one road test.

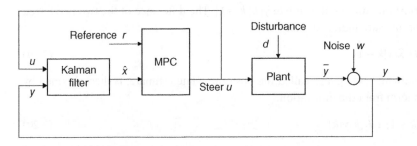

Figure 7.3 MPC framework with Kalman filter for path following.

Equation (7.23) is then discretized and used to update the state and output estimate in the Kalman filter algorithm. The Kalman filter gives a probabilistic state estimate $\hat{\mathbf{x}}_{k|k}$ for time step k by iterations of the discrete Riccati equation and the measurement information [12]. The noise covariance matrices Q_s and Q_m that appear in the Riccati equation are associated with the noises of the model and measurements, respectively, which gives the relative fidelity between the plant model and the measurement.

Figure 7.2 shows the measurement data of yaw rate from the IMU sensor and the Kalman filter estimation results of r and v_y during one vehicle test experiment. The Kalman filter makes the data of vehicle lateral velocity v_y available and ready to use for the MPC controller.

An overall diagram showing the MPC framework along with the Kalman filter is shown in Figure 7.3.

7.2.3 Experimental Results

The proposed linear MPC controller runs at 25 Hz frequency and is evaluated through experiments with our Ford Fusion test vehicle using its online implementation in the on-board dSpace MicroAutoBox controller. The architecture of the Ford Fusion test vehicle used here was presented in detail in Chapter 3. Since the MPC controller takes a future horizon to determine its current control action, it has less stringent requirements of the path description. Therefore, a discrete waypoint representation is used during the test instead of a mathematical function. The use of discrete waypoints also facilitates the later online adjustments for other purposes such as collision avoidance. The elastic-band method is chosen for such collision avoidance waypoint adjustments so that the vehicle can be reactive to obstacles ahead. The methodology of the elastic-band method is not presented in detail in this chapter as it was presented earlier in the book. Its main idea is very straightforward and intuitively easy to understand when the path is viewed as an elastic band subject to virtual forces generated by nearby objects that we want to avoid hitting [13–15]. Two scenarios are chosen to demonstrate the online waypoints adjustment procedure and the MPC path following system performance.

Multiple Static Pedestrians at Parking Lot

In the first scenario, two virtual pedestrians are set with their positions fixed by the aisle to mimic a typical parking lot scenario. The original path is straight through the aisle but would unavoidably collide with the pedestrians. The elastic-band method is able to adjust the waypoints to form a smooth new path as shown in Figure 7.4. The MPC controller is used to follow this modified path and is shown as the vehicle trajectory in Figure 7.4. Note that the resulting vehicle trajectory is very close to the updated path in the plot with a standard deviation of lateral tracking error of around 0.1294 m, which validates the path following performance of the MPC controller used here.

Figure 7.5 shows the sensor measurements of vehicle driving speed, steering wheel angle, and yaw rate during the experiment illustrated in Figure 7.4. The vehicle initial speed is about 2 m/s and gradually increases to about 6 m/s. These low speeds are meaningful in a parking lot where vehicles should be driving slowly. The steering wheel angle determined by the MPC path-following controller and the resultant vehicle yaw rate are observed to be smoothly varying and without oscillations despite the relatively rough calculations of reference position and heading for the discrete waypoint path representation. These smooth steering angle and yaw rate values will be beneficial for maintaining passenger comfort.

Jogging Pedestrian Alongside the Road

The second scenario is more challenging with a jogging pedestrian moving alongside a public road as shown in Figure 7.6. The original discrete waypoint represented path shown with plus signs in Figure 7.6 has to be modified to avoid a possible collision with the pedestrian. The elastic-band

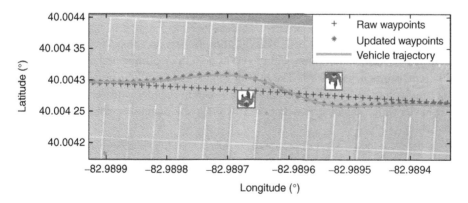

Figure 7.4 Vehicle experiment results showing the waypoints adjustment to avoid virtual pedestrians in a parking lot and the MPC path following trajectory.

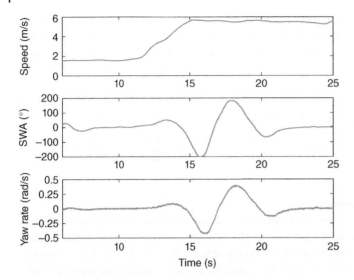

Figure 7.5 Measurement of vehicle speed, steering wheel angle, and yaw rate during the vehicle experiment in Figure 7.4.

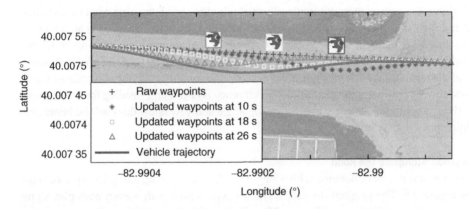

Figure 7.6 Vehicle experiment results showing real-time waypoints adjustment to avoid the virtual moving pedestrian aside the road and the MPC path following trajectory.

method updates the path depending on the pedestrian location at every time step. The virtual pedestrian location at time 10, 18, and 26 seconds, and the corresponding real-time updated waypoints are shown in different shapes in Figure 7.6. The overall vehicle trajectory illustrated with a solid line and the trajectory following the changing path is also seen in the plot with very natural and smooth evasive behavior which demonstrates that the MPC is capable of following the continuously changing path.

Figure 7.7 shows the recorded sensor measurements of speed, steering wheel angle, and yaw rate during the experiment of Figure 7.6. The steering wheel angle is less smooth as compared to the static pedestrian case in Figure 7.5. This is due to the path being updated at every time step by the elastic-band method as the pedestrian is moving. This makes the preview horizon in the MPC control less consistent with the one in the previous time cycle and is reflected in the small oscillations of the MPC determined steering angle in Figure 7.7. Overall, these higher-frequency oscillations

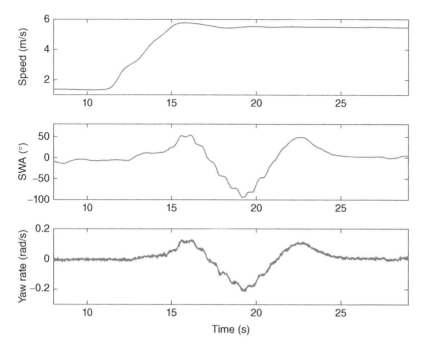

Figure 7.7 Measurement of vehicle speed, steering wheel angle, and yaw rate during the scenario in Figure 7.6.

observed in the steering wheel angle and yaw rate measurements are relatively insignificant and their impact on passenger comfort is limited.

In summary, the experiments with the MPC path-tracking control along with the real-time way-points adjustment for collision avoidance demonstrate effective path following performance as well as the vehicle's reactive capability to avoid potential collision with surrounding objects. The path represented by discrete waypoints shows more flexibility as compared to the spline path for easy adjustments during real-time replanning of the path for collision avoidance. The MPC-path following controller is also favorable owing to its smooth steering action due to its use of a preview horizon and the penalty placed on the rate of change of control in its cost function.

7.3 Design Methodology for Robust Gain-Scheduling Law

7.3.1 Problem Formulation

Plant dynamics usually varies with operating conditions. Gain scheduling is of interest for these plants with a family of scheduled control parameters, each of which provides satisfactory performance for a different operating condition and uses a measurable scheduling variable like vehicle speed. This approach has the advantage of reacting quickly to condition changes as no parameter estimation occurs compared to other types of adaptive controllers. It has been implemented successfully in applications like ship steering [16], process control [17], and especially flight control systems [18, 19] where altitude and Mach number are usually set as scheduling variables. Meanwhile, parametric uncertainties or unstructured dynamics demand the robustness of gain-scheduling control at every operating condition. This demands design methods for a robust gain scheduling control law.

Consider a typical state-space model that describes the plant dynamics as

$$\begin{cases} \dot{\mathbf{x}} = \mathbf{A}(\mathbf{q})\mathbf{x} + \mathbf{B}(\mathbf{q})\mathbf{u} \\ \mathbf{y} = \mathbf{C}(\mathbf{q})\mathbf{x} \end{cases}, \tag{7.27}$$

where the plant states are $\mathbf{x} \in \mathbb{R}^n$, the input is $\mathbf{u} \in \mathbb{R}^p$, and the output is $\mathbf{y} \in \mathbb{R}^q$. $\mathbf{q}$ represents plant parametric uncertainty and is given by the vector

$$\mathbf{q} = [q_1, q_2, \dots, q_\ell]^T \in Q, \tag{7.28}$$

where the operating domain Q of individual uncertainties q_i is typically represented with the hyper-rectangle

$$Q = \left\{ \mathbf{q} | q_i \in [q_i^-, q_i^+], i = 1, 2, \dots, \ell \right\}. \tag{7.29}$$

If the output-feedback control law is assumed to be

$$\mathbf{u} = -\mathbf{K}\mathbf{y} + \mathbf{r}, \tag{7.30}$$

where $\mathbf{r}$ is the reference input and one or more plant parameters $q_i \in \mathbf{q}$ are chosen as gain-scheduling variables, the question arises on how to design appropriate control gain $\mathbf{K}$ with respect to q_i in order to

(i) satisfy desired system performance characteristics, e.g. reference tracking and disturbance attenuation,
(ii) maintain robust stability despite the parametric change of $\mathbf{q}$.

7.3.2 Design via Optimization in Linear Matrix Inequalities Form

Theoretic research work on robust gain-scheduling design is mostly based on uncertain linear parameter-varying (LPV) models. It becomes a natural technique to formulate the gain-scheduling problem in the form of LMIs, which can then be easily addressed with optimization techniques [20–22]. The formulation of LMI constraints is made possible by conversion of the stability condition and $\mathcal{H}_\infty$ (or ℓ_2) performance indices, as well as assumptions of the control structure. Continuous developments using this idea to form proper LMI problems are abundantly available in the literature. In [23–25], for example, parameter-dependent Lyapunov functions instead of fixed ones were exploited to reduce the conservativeness in characterizing stability and performance. In [26, 27], actuator saturation was considered and reflected in the LMI constraints. In [28, 29], the Youla–Kucera parameterization of all stabilizing controllers was exploited and discussed. However, extensive assumptions were made in order to develop LMI formulations which reduced the potential solution searching space. The LMI design also attempts to solve a single optimization under chosen performance criteria instead of graphically showing the feasible solution area.

A general approach to design robust gain-scheduling control is to form an optimization problem with LMIs from the LPV system. Similar to [30, 31], the state-space equation (7.27) is reorganized in the form of the two-polytope LPV representation:

$$\dot{\mathbf{x}} = \widehat{A}(\lambda)\mathbf{x} + B_1(\lambda)\rho_{ref}$$

$$= \sum_{i=1}^{2} \lambda_i [(A_i + \Delta\widetilde{A}_i + B_{2,i}K_iC)\mathbf{x} + B_{1,i}\rho_{ref}], \tag{7.31}$$

Figure 7.8 Region $\mathcal{D}$ defined in the s-plane.

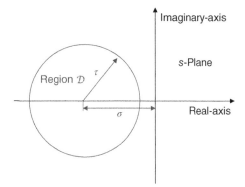

where the gain-scheduling control is assumed to be linear between gains for the two polytopes,

$$u(t) = K(\lambda)Cx(t) = \sum_{i=1}^{2} \lambda_i K_i Cx(t), \tag{7.32}$$

with coefficients satisfying $\lambda_1 + \lambda_2 = 1$, and $\Delta\tilde{A}_i = E_i\tilde{M}(t)F_i$ is the uncertainty matrix due to plant parameters with E_i indicating element wise uncertain range, $\tilde{M}(t) = N(t)I$, $F_i = I$, and scalar $|N(t)| \leq 1$.

A region $\mathcal{D}$ in the s-plane (as illustrated in Figure 7.8) with disk-shape centered at $(-\sigma, 0)$ with radius τ is expressed as follows:

$$\mathcal{D} = \left\{ z \in \mathbb{C} : \begin{bmatrix} -\tau & \sigma + z \\ \sigma + z^* & -\tau \end{bmatrix} < 0 \right\}, \tag{7.33}$$

where we choose $\sigma = 23$ and $\tau = 22.4$ to confine the root set within the covered region similar to what is used later in parameter-space design.

To restrict the response of tracking error from the curvature, the following H_∞ performance indices are formulated:

$$\|\Delta\psi\|_2 < \gamma_1\|\kappa_{ref}\|_2, \tag{7.34}$$

$$\|e\|_2 < \gamma_2\|\kappa_{ref}\|_2. \tag{7.35}$$

From [31], the $\mathcal{D}$-stability (7.33) and H_∞ performance ((7.34)–(7.37)) are guaranteed if $\exists Y_1 > 0$, $Y_2 > 0$, and scalars $\varepsilon_{1,i}$, $\varepsilon_{2,i}$, and $\varepsilon_{3,i}$ such that the followings conditions hold:

$$\Xi_{ij} + \Xi_{ji} < 0, \tag{7.36}$$

$$\Gamma_{1,ij} + \Gamma_{1,ji} < 0, \tag{7.37}$$

$$\Gamma_{2,ij} + \Gamma_{2,ji} < 0, \tag{7.38}$$

where

$$\Xi_{ij} = \begin{bmatrix} -\tau Y_1 & \sigma Y_1 + A_i Y_1 + B_{2,i} K_i C Y_1 & \varepsilon_{1,i} E_j & 0 \\ * & -\tau Y_1 & 0 & (F_i Y_1)^T \\ * & * & -\varepsilon_{1,i} I & 0 \\ * & * & * & -\varepsilon_{1,i} I \end{bmatrix}, \tag{7.39}$$

$$\Gamma_{1,ij} = \begin{bmatrix} \Omega_i & B_{1,i} & Y_2 C_1^T & \varepsilon_{2,i} E_j & Y_2 F_i^T \\ * & -\gamma_1 I & 0 & 0 & 0 \\ * & * & -\gamma_1 I & 0 & 0 \\ * & * & * & -\varepsilon_{2,i} I & 0 \\ * & * & * & * & -\varepsilon_{2,i} I \end{bmatrix}, \tag{7.40}$$

$$\Gamma_{2,ij} = \begin{bmatrix} \Omega_i & B_{1,i} & Y_2 C_2^T & \varepsilon_{3,i} E_j & Y_2 F_i^T \\ * & -\gamma_2 I & 0 & 0 & 0 \\ * & * & -\gamma_2 I & 0 & 0 \\ * & * & * & -\varepsilon_{3,i} I & 0 \\ * & * & * & * & -\varepsilon_{3,i} I \end{bmatrix}, \quad 1 \le i \le j \le 2, \tag{7.41}$$

and $\Omega_i = (A_i Y_2 + B_2 K_i C Y_2) + (A_i Y_2 + B_2 K_i C Y_2)^T$. However, due to the coupling of K_i and positive-definite matrices Y_1 and Y_2, the conditions in (7.39)–(7.41) are bilinear matrix inequalities and are difficult to solve. These conditions can be reduced to LMIs by assuming $Y_1 = Y_2 = Y$ and the substitution of $\widehat{K}_i = K_i CY$. The weighted H_∞ minimization problem

$$\min \gamma_1 + w\gamma_2, \tag{7.42}$$

under the LMI constraints is approachable using semidefinite solver packages such as MOSEK [32].

7.3.3 Parameter-Space Gain-Scheduling Methodology

The parameter-space approach design, on the other hand, gives a set of admissible controls instead of a single "optimal" controller, which gives room for adjusting the trade-offs at the low-level controller design phase. Its graphical representation is also straightforward and easy to interpret for engineering design. The main idea behind the parameter-space approach is that the boundaries of regions of interest in the s-plane or frequency domain cannot be crossed without a continuous change of design parameters. It has previously been implemented in state feedback [33], PID, repetitive control, and disturbance observer design for systems such as the atomic force microscope [34], teleoperation robotics [35, 36], and ground vehicles [37–39], but its application to the gain-scheduling path-tracking controller design is not so common.

Pole Placement and D-Stability

Pole placement for the closed-loop characteristic polynomial is commonly used [40] as an indirect approach to meet time-domain specifications. The closed-loop characteristic polynomial for the system in (7.27) is

$$p(s, \mathbf{q}, \mathbf{k}) = \det[s\mathbf{I} - \mathbf{A}(\mathbf{q}) + \mathbf{B}(\mathbf{q})\mathbf{K}\mathbf{C}(\mathbf{q})], \tag{7.43}$$

where $\mathbf{k}$ is the vector of control gains to be determined in $\mathbf{K}$ given by

$$\mathbf{k} = [k_1, k_2, \dots, k_\tau]^T. \tag{7.44}$$

A more computationally practical way of obtaining the characteristic equation $p(s, \mathbf{q}, \mathbf{k})$ is

$$p(s, \mathbf{q}, \mathbf{k}) = \text{num}\{1 + L(s, \mathbf{q}, \mathbf{k})\}, \tag{7.45}$$

where "num" stands for "numerator of," and the loop gain $L(s, \mathbf{q}, \mathbf{k})$ can be represented linearly with $\mathbf{k}$ by using the Leverrier–Faddeev algorithm [41] on its component $[s\mathbf{I} - \mathbf{A}(\mathbf{q})]^{-1}$.

The set of plant parameter $\mathbf{q}$ or control gains $\mathbf{k}$ or a combination can be used in analysis together in $p(s, \mathbf{q}, \mathbf{k})$ without loss of generality. Therefore, we expand the set $\mathbf{q}$ of uncertainty parameters to inherently include $\mathbf{k}$ as well and denote $p(s, \mathbf{q}, \mathbf{k})$ instead only as $p(s, \mathbf{q})$. The desired region $\mathcal{D}$ for

Figure 7.9 Pole region $\mathcal{D}$ in the s-plane with specifications. Source: Zhu et al. [4] (© [2019] IEEE).

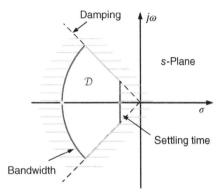

closed-loop pole placement is bounded by its boundary ∂D consisting of one or more contours in the s-plane represented as

$$\partial D := \{s | s = \sigma(\alpha) + j\omega(\alpha), \quad \alpha \in [\alpha^-, \alpha^+]\}. \tag{7.46}$$

Figure 7.9 shows a typical desired pole region $\mathcal{D}$ confined by the limitations of bandwidth, damping, and settling time. Note that the desired pole locations region $\mathcal{D}$ can be disjointed as well.

A polynomial $p(s, \mathbf{q})$ is called $\mathcal{D}$-stable if all its roots $s_i \in \mathcal{D}$. The boundary crossing theorem [40] states that the family of polynomials

$$P(s, Q) = \{a_0(\mathbf{q}) + a_1(\mathbf{q})s + \cdots + a_n(\mathbf{q})s^n | \mathbf{q} \in Q\} \tag{7.47}$$

with continuous real coefficients $a_i(\mathbf{q}) \in \mathbb{R}$ is robustly $\mathcal{D}$-stable, if and only if: (i) there exists a $\mathcal{D}$-stable polynomial $p(s, \mathbf{q}) \in P(s, Q)$, and (ii) $\partial D \cap Roots[P(s, Q)] = \emptyset$, where $Roots[P(s,Q)]$ denotes the set of all roots of $P(s, Q)$. This is the fundamental idea of the parameter space approach. Intuitively, assuming that the real coefficients of $p(s, \mathbf{q})$ are continuous in $\mathbf{q}$, then its roots are also continuous in $\mathbf{q}$, i.e. the roots of $p(s, \mathbf{q})$ cannot jump out of region $\mathcal{D}$ without first crossing its boundary ∂D. Note that there are other more recent approaches to finding the $\mathcal{D}$-stable region by using the Lyapunov equation instead of the boundary crossing theorem [42].

The critical $\mathcal{D}$-stability condition can be mapped to the space of parameters of interest $\mathbf{q}$ by substituting the value of s on the boundary ∂D to $p(s, \mathbf{q})$ via a sweep over generalized frequency α. The boundary obtained in the $\mathbf{q}$-plane is called the complex root boundary (CRB):

$$\partial Q_{CRB}(\alpha) := \{\mathbf{q} | p(\sigma(\alpha) + j\omega(\alpha), \mathbf{q}) = 0, \quad \alpha \in [\alpha^-, \alpha^+]\}. \tag{7.48}$$

This is a complex equation that can be decomposed into two independent equations for its real and imaginary parts which can be solved to decouple two chosen parameters in $\mathbf{q}$ from the $\mathcal{D}$-boundary description. In the special cases, where ∂D crosses the real axis or goes to infinity, the special and degenerate cases of the real root boundary (RRB) and infinite root boundary (IRB) also appear in the $\mathbf{q}$-plane:

$$\partial Q_{RRB} := \{\mathbf{q} | p(\sigma_0, \mathbf{q}) = 0\}, \tag{7.49}$$

$$\partial Q_{IRB} := \{\mathbf{q} | \lim_{\alpha \to \infty} p(\sigma(\alpha) + j\omega(\alpha), \mathbf{q}) = 0\}, \tag{7.50}$$

where σ_0 is the intersection of ∂D on the real axis. The conditions in Eqs. (7.49) and (7.50) reduce to a single equation as compared to the two independent equations of (7.48).

Frequency Response Magnitude (FRM) Specifications

For single input single output (SISO) systems, the robust stability and performance requirements can be clearly represented in the frequency response magnitude (FRM) domain according to the H_∞ control theory. Therefore, similar to the desired pole location region D, we can define a region B in the FRM-plane confined by its boundary $\partial B(\omega)$, $\omega \in [\omega^-, \omega^+]$, where the lower and upper frequency bound ω^- and ω^+ could take any values from 0 to ∞.

The boundary $\partial B(\omega)$ puts different emphasis at various frequencies. An exemplary region B for transfer function magnitude frequency response $|G(j\omega)|$ is illustrated in Figure 7.10.

A transfer function $G(j\omega, Q)$ is called B-stable if its magnitude frequency response lies in region B. Similar to the boundary-crossing theorem in the s-plane, it is derived from the mean value theorem that the family of transfer functions

$$\mathcal{G}(j\omega, Q) = \{G(j\omega, \mathbf{q} | \mathbf{q} \in Q\}, \tag{7.51}$$

is B-stable if and only if: (i) There exists a B-stable transfer function $G(j\omega, q) \in \mathcal{G}(j\omega, q)$, and (ii) $|\mathcal{G}(j\omega, Q)| \neq \partial B(\omega)$, $\forall \omega \in [\omega^-, \omega^+]$ [43].

Points on ∂B can be mapped to the boundary ∂Q_B in the **q**-plane by a sweep over frequency ω:

$$\partial Q_B := \{ \mathbf{q} | |G(j\omega, \mathbf{q})| = \partial B(\omega), \forall \omega \in [\omega^-, \omega^+]\}. \tag{7.52}$$

At each fixed frequency ω^*, this generally forms an implicit equation of the uncertain parameters **q**. Boundary curves may be plotted by gridding some parameters in **q** and solving for the rest. In some special cases, this point condition to visualize ∂Q_B in the **q**-plane can be simplified. For example, H_∞ robust performance is ensured by satisfying the mixed sensitivity constraint

$$\| |W_S(s)S(s, \mathbf{q})| + |W_T(s)T(s, \mathbf{q})| \|_\infty < 1, \tag{7.53}$$

where W_S is the weight for sensitivity function S and depends on nominal performance requirements like disturbance rejection or reference tracking, W_T is the weight for the complementary sensitivity function T and is usually set to be the multiplicative uncertainty bound of the plant, and the boundary in the FRM domain here is a constant $\partial B = 1$.

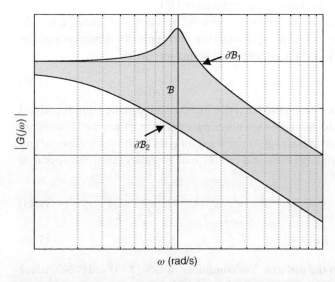

Figure 7.10 An exemplary region B for transfer function $G(s)$. Source: Zhu et al. [4] (© [2019] IEEE).

Figure 7.11 Illustration of critical condition of mixed sensitivity constraint in Nyquist plane. Source: Zhu et al. [4] (© [2019] IEEE).

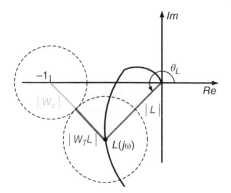

Equation (7.53) can be rewritten as

$$\left| \frac{W_S(j\omega)}{1 + L(j\omega, \mathbf{q})} \right| + \left| \frac{W_T(j\omega)L(j\omega, \mathbf{q})}{1 + L(j\omega, \mathbf{q})} \right| < 1, \tag{7.54}$$

or

$$|W_S(j\omega)| + |W_T(j\omega)L(j\omega, \mathbf{q})| < |1 + L(j\omega, \mathbf{q})|, \quad \forall \omega. \tag{7.55}$$

The critical condition happens at some frequency ω where the strict inequality in Eq. (7.54) becomes an equality. The graphical illustration of this condition could be represented in the Nyquist plane (Figure 7.11) and forms the shaded triangle. Applying the cosine rule to this triangle, a graphical solution for $|L|$ may be obtained [44]:

$$|L| = \frac{-\cos(\theta_L) + |W_S||W_T| \pm \sqrt{Disc}}{1 - |W_T|^2}, \tag{7.56}$$

where

$$Disc = 1 + \cos^2 \theta_L - 2|W_S||W_T| \cos \theta_L + |W_S|^2 + |W_T|^2. \tag{7.57}$$

The solution procedure is to sweep the angle θ_L from 0 to 2π rad, and evaluate Eq. (7.56) at the chosen frequency ω to see if any solutions exist for $|L|$ at each value of θ_L. Then, all possible solutions of $L(j\omega, \mathbf{q}) = |L|e^{j\theta_L}$ at frequency ω can be obtained, and the relationship within $\mathbf{q}$ is obtained by solving the two equations, real and imaginary, within the complex valued equation

$$L(j\omega, \mathbf{q}) = Re(L) + j \, Im(L) \tag{7.58}$$

for two chosen parameters in $\mathbf{q}$. Repeating this procedure at grid points in the frequency range $[\omega^-, \omega^+]$, the boundary $\partial Q_\mathcal{B}$ in the $\mathbf{q}$-plane can be gradually formed for the mixed sensitivity constraint. It is possible to repeat this procedure for different values of a third chosen variable in $\mathbf{q}$, thereby obtaining a three-dimensional parameter space representation.

Robust Gain-Scheduling Design Procedure

To visualize these boundaries in the Euclidean space, usually up to three parameters of interest $\mathbf{q}^* \subset \mathbf{q}$ are chosen with its complementary set of parameters $\overline{\mathbf{q}}^*$ fixed. For the gain-scheduling design in our case, $\mathbf{q}^*$ should include at least one plant parameter as the scheduling variable, as well as control gains. Then applying Eqs. (7.48)–(7.52) plus (7.52) and holding other parameters $\overline{\mathbf{q}}^*$ fixed, the boundaries ∂D and ∂B can be visualized in the $\mathbf{q}^*$-plane. In this manner, the region where parameters $\mathbf{q}^*$ ensure D-stability and $\mathcal{B}$-stability at the fixed operating condition $\overline{\mathbf{q}}^*$ is obtained.

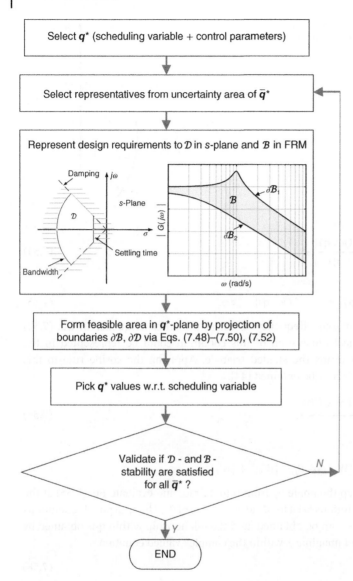

Figure 7.12 Parameter-space based robust gain-scheduling design diagram. Source: Zhu et al. [4] (© [2019] IEEE).

The parameter-space-based robust gain-scheduling design process is summarized in Figure 7.12 and detailed below:

(i) Select parameters of interest **q***, including scheduling variable and control parameter(s).

(ii) Select representatives, i.e. different fixed operating conditions $\bar{\mathbf{q}}$*, from the uncertainty area Q. Vertices or vertices and some other points on the boundary Q are usually are chosen as representatives to include the extreme cases into design consideration.

(iii) Represent design requirements in the form of region D in the s-plane and region B in the FRM domain. The boundaries ∂D and ∂B are then projected to the **q***-plane at each representative via ((7.48)–(7.52)) and (7.52). The feasible region confined by these projected boundaries in the

q*-plane is where $\mathcal{D}$-stability and $\mathcal{B}$-stability are ensured at the representative. Superimpose these feasible regions at different representatives altogether in the **q***-plane.

(iv) If an overlapped feasible region exists, pick parameters **q*** from inside, i.e. control gains corresponding to the scheduling variable. Then, $\mathcal{D}$-stability and $\mathcal{B}$-stability are both satisfied at the chosen values of **q*** for all representatives.

(v) Validate the $\mathcal{D}$-stability and $\mathcal{B}$-stability at the selected **q*** for all possible operating conditions $\overline{\mathbf{q}}^*$ in the uncertainty area by gridding. An alternative for validation is to map the boundaries $\partial \mathcal{D}$ and $\partial \mathcal{B}$ at the selected **q*** into the $\overline{\mathbf{q}}^*$-plane and check if the uncertainty area is bounded inside. If validation fails, go back to step (iii), add more representatives and pick new **q*** until (v) is satisfied.

7.4 Robust Gain-Scheduling Application to Path-Tracking Control

7.4.1 Car Steering Model and Parameter Uncertainty

The robust gain-scheduling controller design method presented in Section 7.2 is applied to automated vehicle lateral control in this section. The vehicle is required to track a planned path while maintaining robust stability in the presence of changes in operating condition and uncertainties of vehicle load and tire saturation. Gain scheduling through the use of a speed-dependent look-ahead distance is used for vehicle lateral control [45–47]. This is similar to what a driver automatically does by visually previewing the road ahead using speed-dependent preview distance. In this section, the scheduled gain is selected from the space of design parameters where stability and robust performance requirements were projected. The analysis is conducted for the whole uncertainty region and further validated through HIL simulations and road tests with our experimental vehicle. A comparative study with the previously presented LMI steering controller design is carried out and presented in the form of discussions and experimental results.

The single-track car model as illustrated in Figure 7.13 provides an accurate description of vehicle lateral dynamics and is customarily used in steering controller design. The single-track car model becomes linearized if the lateral force of the tire is assumed proportional to its sideslip angle which simplifies the steering controller design. However, this assumption is invalid if the tire slip angle becomes large and correspondingly leads to worsened model accuracy.

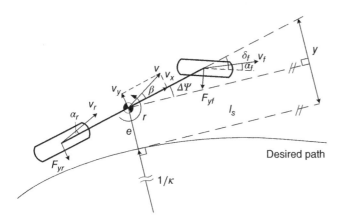

Figure 7.13 Single-track car model and its deviation from the desired path.

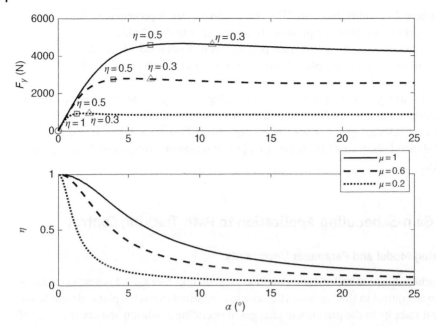

Figure 7.14 Tire lateral forces F_y and tire saturation parameter η with respect to slip angle at different road friction coefficient. Source: Zhu et al. [4] (© [2019] IEEE).

To include the tire nonlinearities in the robustness analysis, the tire saturation parameter η [48, 49] is introduced as a factor in that modifies the tire lateral force F_y given by

$$F_y = \eta C_\alpha \alpha, \tag{7.59}$$

where C_α is the nominal value of the tire cornering stiffness at small tire slip angle. Typical lateral forces of a tire (215/55R17) under load (4780 N) with respect to tire slip angle α at different road friction coefficients μ are shown in Figure 7.14. Values of $\eta = 1, 0.5,$ and 0.3 are marked on the lateral force curves. The introduced tire saturation parameter η is seen to be descending with increasing tire slip angle α, and limited by the road friction coefficient μ.

With the introduced tire saturation parameter η, the LPV model that describes the vehicle states and their deviation from the planned path in state-space form is (see derivation in the appendix of [4]):

$$
\begin{bmatrix} \dot{\beta} \\ \dot{r} \\ \Delta\dot{\psi} \\ \dot{y} \end{bmatrix} =
\begin{bmatrix} a_{11} & a_{12} & 0 & 0 \\ a_{21} & a_{22} & 0 & 0 \\ 0 & 1 & 0 & 0 \\ v_x & l_s & v_x & 0 \end{bmatrix}
\begin{bmatrix} \beta \\ r \\ \Delta\psi \\ y \end{bmatrix} +
\begin{bmatrix} b_{11} & 0 \\ b_{21} & 0 \\ 0 & v_x \\ 0 & -l_s v_x \end{bmatrix}
\begin{bmatrix} \delta_f \\ \kappa_{ref} \end{bmatrix}, \tag{7.60}
$$

with coefficients

$$a_{11} = -\frac{C_{af} + C_{ar}}{(m/\eta)v_x}, \tag{7.61}$$

$$a_{12} = -1 - \frac{C_{af}l_f - C_{ar}l_r}{(m/\eta)v_x^2}, \tag{7.62}$$

$$a_{21} = -\frac{C_{af}l_f - C_{ar}l_r}{(I_z/\eta)}, \tag{7.63}$$

$$a_{22} = -\frac{C_{af}l_f^2 + C_{ar}l_r^2}{(I_z/\eta)v_x}, \tag{7.64}$$

$$b_{11} = \frac{C_{af}}{(m/\eta)v_x}, \tag{7.65}$$

$$b_{21} = \frac{C_{af}l_f}{(I_z/\eta)}, \tag{7.66}$$

where C_{af}, C_{ar} are lumped tire cornering stiffnesses for front and rear tires, respectively; v_x is the vehicle longitudinal speed; l_f, l_r are the lengths from the vehicle center of gravity (CoG) to its front and rear axles, respectively; β is the vehicle side slip angle; r is the yaw rate of the vehicle; $\Delta\psi$ is the yaw angle error from the path; y is the deviation of the vehicle from the path at the look-ahead distance l_s (Figure 7.13); δ_f is the steering angle of the front wheel; κ_{ref} is the path reference curvature.

The uncertainty range of plant parameters, m, v_x, and η were determined by considering different vehicle operating conditions: $m \in [1700, 2060]$ kg, $v_x \in [5, 30]$ m/s. The range of the tire saturation parameter η is selected empirically as $[0.2(0.1v_x + 1), 1]$, with its bottom limit increasing with speed. The reason behind this selection is that tire force saturation at high speed is more dangerous than at low speed. The vehicle is prone to losing its maneuverability and may not be stabilized by steering control alone, at which point more actuators such as brake control would be necessary to intervene. Therefore, we consider that it is adequate to tighten the range of η as speed increases during the steering control design phase.

The parametric uncertainty region is visualized as shown in Figure 7.15. Note that in the model (7.60), coefficients m and η are lumped together as m/η, the extreme value of which is obtained at either (m^-, η^+) or (m^+, η^-). The model representatives are, thus, chosen as these extremes within a grid of speed values v_x as indicated with the dots on the leftmost line in Figure 7.15.

7.4.2 Controller Structure and Design Parameters

The system diagram of the proposed controller along with the plant is shown in Figure 7.16, where the control u is the front-wheel steering angle δ_f, the output y is the look-ahead error, and n is the

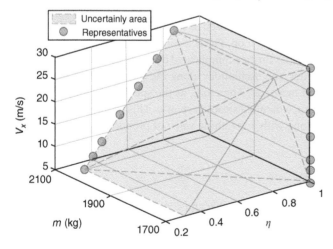

Figure 7.15 Uncertainty region of plant parameters and selected representatives. Source: Zhu et al. [4] (© [2019] IEEE).

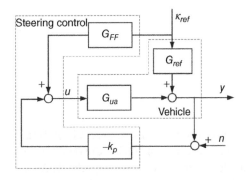

sensor noise. The transfer functions, G_{ua} and G_{ref}, from u and κ_{ref} to y, respectively, were derived from the state-space representation in (7.60). G_{ref} has the form as

$$\frac{Y(s)}{\kappa_{ref}(s)} = G_{ref}(s) = -\frac{l_s v_x (s + v_x/l_s)}{s^2}. \tag{7.67}$$

The controller structure is comprised of a feedforward loop G_{FF} which adds the required steady-state steering angle for following curvature κ_{ref} to the input along with a proportional feedback control loop to reduce the remaining tracking error y. The feedforward loop gain is

$$G_{FF} = L + K_{us} v_x^2, \tag{7.68}$$

which incorporates the approximation of steady-state steering gain for following a curvature. K_{us} is called the understeer gradient and is given by

$$K_{us} = \frac{m_0}{L} \left(\frac{l_r}{C_{af}} - \frac{l_f}{C_{ar}} \right), \tag{7.69}$$

where the wheelbase $L = l_f + l_r = 2.8\,\text{m}$ and a nominal vehicle mass $m_0 = 1850\,\text{kg}$ are used.

The proportional feedback is

$$-k_p y = -k_p [e + l_s \, \sin(\Delta \psi)], \tag{7.70}$$

where e is the lateral error of vehicle CoG from the path (Figure 7.13). Its robust stability in regions of high tire saturations at various speeds is proved in [49] using the Lyapunov method.

The design parameters of the steering controller are chosen as the proportional gain k_p and the look-ahead distance l_s in the proportional feedback loop. The selection of l_s is a trade-off between rapid accuracy and damped response [45], and is further complicated by the variation of the vehicle speed. If the set of k_p and l_s are fixed values, the oscillations along the path could become high frequency and intolerable when the speed increases. Therefore, as the speed rises, an increasing look-ahead distance is preferred to maintain a damped steady response. This raises the demand of a gain-scheduling look-ahead distance, along with a matching value of proportional gain at each speed.

Altogether we have the parameter set

$$\mathbf{q} = [m, \eta, v_x, l_s, k_p]^T, \tag{7.71}$$

with the parameters of interest including scheduling variable and control parameters being

$$\mathbf{q}^* = [v_x, l_s, k_p]^T. \tag{7.72}$$

The design for the desired speed-scheduled profiles of both l_s and k_p is not evident and traditionally relies heavily on empirical tuning through slalom tests [45]. Meanwhile, robustness of the controller to model uncertainties should also be guaranteed. The application of the parameter-space approach provides an alternative to satisfying these design requirements for the robust gain-scheduling design.

7.4.3 Application of Parameter-Space Gain-Scheduling

D-Stability

For D-stability design of the automated vehicle lateral control, the pole region D illustrated in Figure 7.9 is used with the design requirements: damping ratio $\zeta > 0.52$, approximate bandwidth limitation of 70 rad/s, and settling time $T < 8$ seconds. At each representative, the boundary ∂D of the D-stable region is projected to the plane of (k_p, l_s) through Eqs. (7.48)–(7.52).

Mixed Sensitivity Constraint

Robust stability of the controller is one consideration as the plant model of vehicle lateral dynamics is subjected to parametric uncertainties due to parameter variations and unstructured uncertainties due to higher frequency unmodeled dynamics. Its multiplicative uncertainty bound is chosen as a first-order system:

$$W_T(s) = h_T \frac{s + \omega_T l_T}{s + \omega_T h_T}, \tag{7.73}$$

with the low-frequency gain $l_T = 0.15$, the high-frequency gain $h_T = 1.0$ (corresponds to up to 100% uncertainty at high frequencies), and the frequency of transition to significant model uncertainty is $\omega_T = 70$ rad/s.

Meanwhile, the road curvature κ_{ref} could be viewed as a disturbance acting on the system (Figure 7.16). Hence, disturbance rejection at the output should also be considered. For simplicity, assuming a step input from the disturbance, the weight for sensitivity function is set as follows:

$$W_S(s) = \frac{1}{s}. \tag{7.74}$$

The robustness of performance is guaranteed if the mixed sensitivity constraint in Eqs. (7.54)–(7.57) can be satisfied. In this case, points on the boundary ∂B can be converted to the implicit equation in q given by (7.58) by a sweep over frequency $\omega \in [\omega^-, \omega^+]$ and $\theta_L \in [0, 2\pi)$ at representative operating conditions $\overline{q}^* = [m, \eta]^T$ as

$$k_p \cdot G_{ua}(j\omega, \overline{q}^*, v_x, l_s) = Re(L) + j\text{Im}(L). \tag{7.75}$$

The plant model $G_{ua}(j\omega, \overline{q}^*, v_x, l_s)$ has the form:

$$G_{ua}(j\omega, \overline{q}^*, v_x, l_s) = c_1(j\omega, \overline{q}^*, V_x) \cdot l_s + c_2(j\omega, \overline{q}^*, v_x), \tag{7.76}$$

where $c_1, c_2 \in \mathbb{C}$ are coefficients decided by ω, $\overline{q}^*$, and v_x, and can be represented in complex form as

$$c_1 = u_1 + jv_1 \quad \text{and} \quad c_2 = u_2 + jv_2, u_1, v_1, u_2, v_2 \in \mathbb{R}. \tag{7.77}$$

Substituting Eqs. (7.76) and (7.77) to Eq. (7.75) and reorganizing results in

$$k_p \cdot [(u_1 l_s + u_2) + j(v_1 l_s + v_2)] = Re(L) + j\text{Im}(L). \tag{7.78}$$

Since $k_p \in \mathbb{R}^+$, from the above equation, we have

$$\frac{u_1 l_s + u_2}{Re(L)} = \frac{v_1 l_s + v_2}{\text{Im}(L)}. \tag{7.79}$$

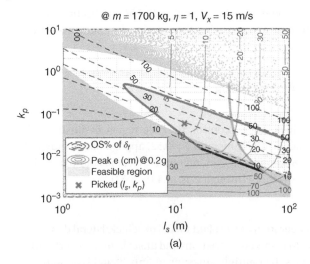

@ $m = 1700$ kg, $\eta = 1$, $V_x = 15$ m/s

l_s (m)

(a)

Figure 7.17 Mappings of boundaries $\partial \mathcal{D}$, mixed sensitivity point conditions, and performance contours in the (l_s, k_p)-plane of two representatives at $v_x = 15$ m/s. Source: Zhu et al. [4] (© [2019] IEEE).

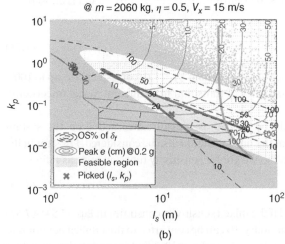

@ $m = 2060$ kg, $\eta = 0.5$, $V_x = 15$ m/s

l_s (m)

(b)

The value of l_s can be obtained from this equation at each sweep point, where ω, θ_L, $\overline{\boldsymbol{q}}^*$, and v_x are held fixed. Once l_s is determined, the control gain k_p can be obtained from Eq. (7.79).

Figure 7.17 shows the mapping of $\mathcal{D}$-stability boundaries to the (l_s, k_p)-plane (in bold solid lines with tones matching $\partial \mathcal{D}$ in the s-plane in Figure 7.9) at two representatives at $v_x = 15$ m/s, as well as points projected from the mixed sensitivity critical condition (in gray dots). It can be observed that these points form an envelope in the plot. The region contained within both the envelope and $\mathcal{D}$-stability boundaries is the feasible region (highlighted in light shade) and control parameters (l_s, k_p) chosen from this feasible region satisfy both $\mathcal{D}$-stability and the mixed sensitivity constraint for this representative.

Performance Contours

Another potential advantage of using the parameter-space approach is that lines of constant system performance measures can easily be calculated and displayed in the chosen plane of design parameters. Given a step input of road curvature with 0.2g steady-state lateral acceleration at v_x,

$$\kappa_{ref}(t) = \begin{cases} 0.2g/v_x^2, & t \geq 0 \\ 0, & t < 0 \end{cases}. \tag{7.80}$$

The peak lateral error e and the overshoot (OS%) of steering δ_f can be obtained by simulation of the linear model in Eq. (7.60) at values of parameter set $\boldsymbol{q}$. The contours of these performance indices can then be formed and displayed as shown in Figure 7.17. The control parameters (l_s, k_p) indicated in cross in the plot were selected at the overlapped feasible regions of the two representatives by also considering the performance contours in an attempt to achieve small error e and small steering overshoot at the same time.

Similarly, the feasible regions in the (l_s, k_p)-plane at other speeds v_x were obtained and the corresponding control parameters were determined as shown in Figure 7.18. These regions altogether form a three-dimensional stack of feasible areas within which the parameters of interest $\boldsymbol{q}^*$ satisfy both $\mathcal{D}$-stability and the mixed sensitivity constraint.

The gain-scheduling control law is hence formed by connecting these different controller gains for each different longitudinal speed value. The scheduled values for l_s and k_p with respect to the scheduling variable v_x are shown in Figure 7.19.

Validation and Analysis

The gain-scheduling control presented here guarantees the $\mathcal{D}$-stability and the robust performance only at the representatives of the uncertain model family. A postdesign robustness analysis is, therefore, necessary to validate whether the same conclusion can be made for the whole uncertainty

Figure 7.18 Feasible regions and selected control parameters (l_s, k_p) at each speed. Source: Zhu et al. [4] (© [2019] IEEE).

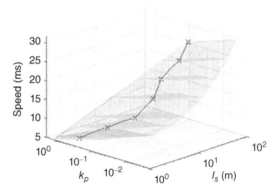

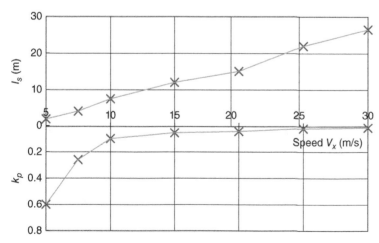

Figure 7.19 Gain scheduling control law with control parameters (l_s, k_p) for each speed. Source: Zhu et al. [4] (© [2019] IEEE).

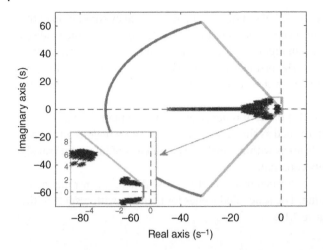

Figure 7.20 Root set for the designed gain-scheduling control. Source: Zhu et al. [4] (© [2019] IEEE).

region of plant parameters in Figure 7.15. This is done through root set analysis. The root set contains the roots of the closed-loop characteristic equation subjected to all possible sets of parameters in the Q-domain:

$$\text{Roots}[P(s, Q)] = \{s | p(s, \mathbf{q}) = 0, \quad \forall \mathbf{q} \in Q\}. \tag{7.81}$$

Figure 7.20 shows the root set for the designed gain-scheduling control which was constructed by gridding parameters $\mathbf{q}$ and solving roots via Eq. (7.45). This plot shows that the roots of all possible operating conditions are within the region $\mathcal{D}$ in the s-plane.

To validate the mixed sensitivity constraint for all operating conditions, the uncertainty bound has to be determined first. The additive parametric uncertainty Δ for the plant is

$$\Delta = G_{ua}(j\omega, \mathbf{q}) - G_{ua}(j\omega, \mathbf{q}_0), \tag{7.82}$$

where G_{ua} is the plant transfer functions, and $\mathbf{q}_0$ is the nominal condition which is usually chosen at the midpoint values of parametric uncertainty bounds. The magnitude frequency responses of $|\Delta|$ at speed $v_x = 15\,\text{m/s}$ by a sweep over $m \in [1700, 2060]$ kg and $\eta \in [0.2(0.1v_x + 1), 1]$ are shown in Figure 7.21. An additive uncertainty bound W_Δ can then be determined by imposing that all $|\Delta|$ plots lie beneath it along the frequency range of interest as

$$W_\Delta = \frac{12(s + 1)}{s^2(0.05s + 1)}. \tag{7.83}$$

Then, the multiplicative uncertainty bound, $W_T = W_\Delta / G_{ua}(j\omega, \mathbf{q}_0)$, along with $\mathbf{q}_0$ and the control gains (l_s, k_p) with respect to v_x are substituted into the left-hand expression in (7.53) at each speed. The frequency responses $|W_S(j\omega)S(j\omega, \mathbf{q}_0)| + |W_T(j\omega)T(j\omega, \mathbf{q}_0)|$ at different speeds are shown in Figure 7.22, revealing that the mixed sensitivity constraint in (7.54) is fulfilled.

7.4.4 Comparative Study of LMI Design

The methodology to form a LMI-constrained optimization problem for robust gain-scheduled control design was covered in Section 7.3.2. To make comparisons with control gains from the

Figure 7.21 Additive parametric uncertainty at $v_x = 15$ m/s and its bound. Source: Zhu et al. [4] (© [2019] IEEE).

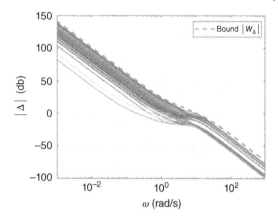

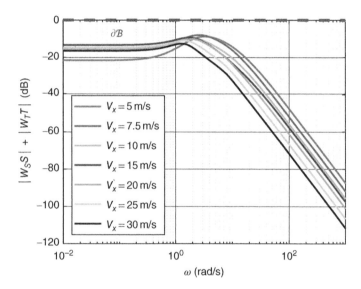

Figure 7.22 Mixed sensitivity constraint validated at different speeds. Source: Zhu et al. [4] (© [2019] IEEE).

parameter-space design, the states for the LMI design are chosen as $\mathbf{x} = [\beta, r, \Delta\psi, e]^T$. Initial attempts to only use the last two states for feedback were tried by assignments of zeros inside Y and $\widehat{K}_i$ to ensure the expression,

$$K_i C = \widehat{K}_i Y^{-1},\tag{7.84}$$

to have zeros for its first two elements. Solutions were not found; however, due to reduced number of free parameters to search under the already-stringent assumptions. Therefore, one more state, r was used as well for feedback. Matrices used were

$$Y = \begin{bmatrix} y_{11} & 0 & 0 & 0 \\ 0 & y_{22} & y_{23} & y_{24} \\ 0 & * & y_{33} & y_{34} \\ 0 & * & * & y_{44} \end{bmatrix}, \quad \widehat{K}_i = \begin{bmatrix} 0 & k_{i2} & k_{i3} & k_{i4} \end{bmatrix}.\tag{7.85}$$

The nonzero state feedback gains were, then, found after semidefinite programming followed by conversion of Eq. (7.84) as follows:

$$K_1 = \begin{bmatrix} -0.4428 & -8.3594 & -0.8994 \end{bmatrix}^T, \tag{7.86}$$

$$K_2 = \begin{bmatrix} -0.0993 & -4.5141 & -0.4876 \end{bmatrix}^T, \tag{7.87}$$

at the two polytopes $v_x = 10$ and $30\,\text{m/s}$. In this process, only parametric uncertainty of tire saturation parameters was considered due to the difficulty of having proper LPV and LMI forms when more uncertain parameters are used.

Different limitations of both approaches were observed during their respective design processes. LMI design is more systematic but may fail to reach solutions due to conservativeness being accumulated due to assumptions made during the formation of LPV and LMI structures, e.g. assumptions $Y_1 = Y_2$ made in (7.36)–(7.40) to form LMI constraints. LMI design also requires an analytical expression of the gain-scheduling control law which limits the potential of the searching scope. The parameter-space design, on the other hand, is intuitive and provides a feasible area for engineers to choose from instead of resulting on just one single solution. It is also less conservative due to the fewer assumptions being made during the design. It may, however, require tedious drawings as the number of parameters of interest $\mathbf{q}^*$ increases. Further road experiments with both parameter-space and LMI designed control laws are later presented in Figures 7.29 and 7.30 with discussions.

7.4.5 Experimental Results and Discussions

Double-Lane Change Path

A double-lane change (DLC) path was chosen to evaluate both the robustness and tracking performance of the vehicle lateral controllers designed in this chapter. The vehicle trajectories are simulated with the HIL simulator shown in Figure 7.23 at different operating conditions. The HIL equipment includes a dSpace SCALEXIO real-time system and a dSpace MicroAutoBox prototype controller. The former runs a high-fidelity CarSim vehicle model with parameters and soft sensors matching our Ford Fusion test vehicle while the MicroAutoBox runs the control algorithm at 100 Hz frequency. The two communicate via the controller area network (CAN) bus, the same as the protocol used in our experimental vehicle for steering actuator control.

The reference path is shown with a dash-dot line in Figure 7.24 modeled by a polynomial spline for the continuity and smoothness in its reference position and heading. The simulation responses

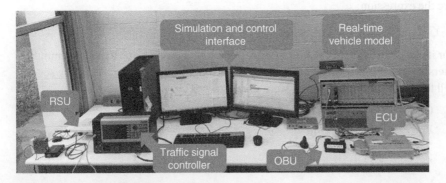

Figure 7.23 Hardware-in-the-loop equipment setup.

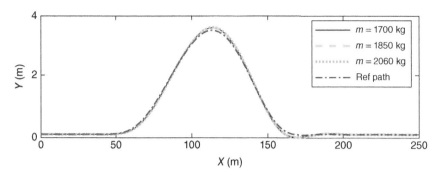

Figure 7.24 Tracking trajectories with different vehicle masses at 20 m/s speed. Source: Zhu et al. [4] (© [2019] IEEE).

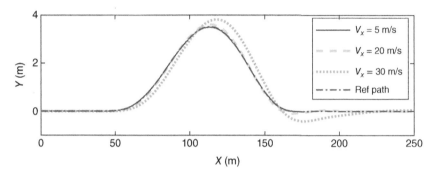

Figure 7.25 Tracking trajectories at different speeds with vehicle mass 1850 kg. Source: Zhu et al. [4] (© [2019] IEEE).

in Figure 7.24 demonstrate the tracking trajectories of vehicles of different mass within the designed mass uncertainty at the longitudinal speed of 20 m/s. The tracking performance was almost unaffected by the mass difference with the designed controller. Similarly, at different speeds, the vehicle was able to maintain stability and achieve acceptable tracking error within 1 m (Figure 7.25). However, as speed increases the vehicle trajectory tends to deviate from the desired path. This can be explained by the increased gain from reference curvature to the look-ahead error as speed rises as seen in Eq. (7.67) and a lowered scheduling gain k_p being required to maintain robust stability.

A comparison of the designed gain-scheduling control with fixed-gain control is also made in Figure 7.26. The fixed-gain control parameters were held fixed at $l_s = 2$ and $k_p = 0.6$, the same as the gain-scheduling control at low speed 5 m/s. However, this control is prone to cause instability as speed increases due to its high gain. The vehicle under this control eventually lost its stability in the attempt to steer back despite its initially accurate tracking as shown in Figure 7.26 at $v_x = 30$ m/s.

The steering wheel angle of the fixed gain control shows a more aggressive behavior (Figure 7.27), which eventually leads to high tire saturation, as indicated by the spiking vehicle sideslip angle. This comparison demonstrates the trade-offs between the tracking accuracy and robust stability. It reveals the necessity of using a gain-scheduling design for the vehicle lateral control.

Slalom Path at Varying Speed

To further evaluate the designed gain-scheduling control at varying scheduling variable, i.e. vehicle speed, we conducted tests along a 200 m road with our experimental vehicle (Figure 7.28). The vehicle is equipped with a dual-antenna global positioning system (GPS) (OXTS xNAV550) with real-time kinematic (RTK) corrections for localization and a steer-by-wire system. The vehicle mass

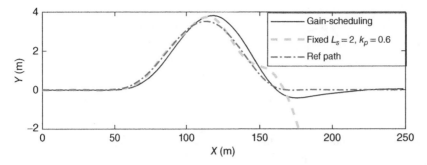

Figure 7.26 Trajectory comparison of gain-scheduling versus fixed gain control at 30 m/s. Source: Zhu et al. [4] (© [2019] IEEE).

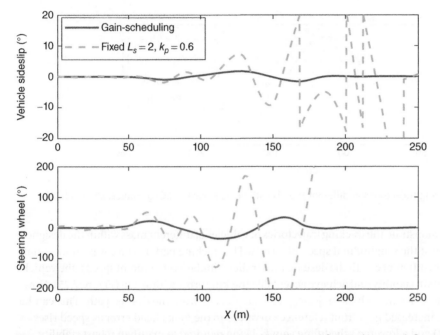

Figure 7.27 Vehicle sideslip and steering of gain-scheduling and fixed gain control. Source: Zhu et al. [4] (© [2019] IEEE).

was estimated around 1865 kg when occupied by two people and test equipment. The longitudinal speed was intentionally set as in Figure 7.29 with dramatic acceleration (up to 9 m/s^2) to evaluate the robustness and adaptivity of the gain-scheduling control. The reference spline-represented path (Figure 7.30) was designed as a sinusoidal slalom type one to better observe the tracking performance. The same speed profile and test during the road experiments were also used in HIL simulation with results matching road tests, as shown in Figures 7.29 and 7.30.

It is observed in Figure 7.30 that tracking accuracy is maintained with the gain-scheduling control despite the fast changes of the scheduling variable, i.e. vehicle speed. At some sections, e.g. 60–80 m, slightly larger deviation from the path was seen. This is probably caused by weakening tire lateral forces due to high tire force saturation at aggressive deceleration as observed in Figure 7.29.

The gain scheduling steering control designed with LMIs in Section IV was also experimented with and results plotted on Figures 7.29 and 7.30. Table 7.1 shows the mean squared tracking error (MSTE) under the three cases with the following definitions of MSTE for orientation and

Figure 7.28 Experimental vehicle on the test road. Source: Zhu et al. [4] (© [2019] IEEE).

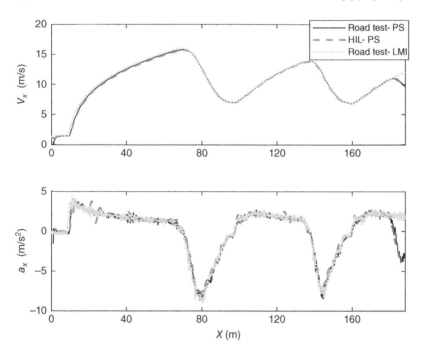

Figure 7.29 Longitudinal speed and acceleration of road test and HIL simulation ("PS" – Parameter-Space design, "LMI" – Linear-matrix-inequality design). Source: Zhu et al. [4] (© [2019] IEEE).

path-tracking error:

$$\sigma_{\Delta\psi} = \sqrt{\frac{\int_0^T |\Delta\psi|^2}{T}} \quad \text{and} \quad \sigma_e = \sqrt{\frac{\int_0^T |e|^2}{T}}. \tag{7.88}$$

It was seen that slightly better-tracking performance at varying speed was achieved by the parameter-space method in comparison to the LMI approach. This improvement may be explained by the introduction of performance contours in the parameter-space design process that can be

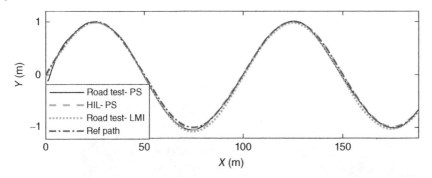

Figure 7.30 Vehicle trajectory at the slalom path ("PS" – Parameter-Space design, "LMI" – Linear-matrix-inequality design). Source: Zhu et al. [4] (© [2019] IEEE).

Table 7.1 Comparisons of tracking errors during experiments.

MSTE	Road test – PS	HIL – PS	Road test – LMI
$\sigma_{\Delta\psi}$ (°)	0.245	0.164	0.264
σ_e (m)	0.029	0.014	0.052

used to reduce the peak lateral error and the steering overshoot. Overall, the parameter-space approach shows its practical value as an alternative to the LMI-constrained optimization method in designing the robust gain-scheduling control law.

7.5 Add-on Vehicle Stability Control for Autonomous Driving

The integration of the road friction coefficient estimation module in Section 3.6 greatly enhances the safety of autonomous driving under adverse road conditions. However, we realized that even if these modules work faultlessly to fulfill their intended functionalities, potential hazardous risks from the complex environment are still of concern for the safety of the intended functionality (SOTIF) defined in ISO/PAS 21448 [50]. For example, in the case when the vehicle suddenly enters a slippery icy road section, the estimation module may not be responsive enough as it requires time and sufficient excitations (acceleration/deceleration movement) to converge to a correct estimation of road friction coefficient. This may lead to an infeasible planned trajectory and a potentially dangerous execution of steering control for path tracking.

It is, hence, beneficial to introduce VSC to reduce the risk of loss of vehicle stability under such occasions. VSC is not a new topic and has long been proposed and used in series production vehicles to deal with the danger of losing vehicle control or stability in versatile cases, such as over-reacted steering maneuver at high speed, driving on slippery icy road in winter, or being suddenly subjected to strong cross-wind. During these critical conditions, it is found that the vehicle steerability, i.e. the capability to provide sufficient yaw moment through front steering, starts to vanish as the vehicle sideslip angle increases [51]. The extra yaw moment required in such circumstances needs to be provided in some form other than steering to maintain or regain the vehicle yaw stability. This naturally necessitates the utilization of tire longitudinal forces from individual wheels to provide that extra yaw moment and forms the main idea of VSC. The VSC system, with other names such as electronic stability control (ESC) or electronic stability program (ESP), has evolved into a mature, series produced and available technology over the last decades. Reports from National Highway Traffic Safety Association (NHTSA) show great effectiveness of VSC in reducing serious accidents, with up to a 70% reduction in fatal rollovers and 36% of fatal single-vehicle crashes for passenger cars [52], the benefits of which lead to government regulations requiring original equipment manufacturers (OEMs) to equip all new passenger vehicles with VSC. In this section, we present and use two different strategies to determine the amount of extra yaw moment that is required, followed by differential braking distribution and simulation results.

7.5.1 Direct Yaw Moment Control Strategies

Yaw Rate-Based Stability Control
The first strategy to decide the amount of extra yaw moment is to estimate an add-on extra yaw moment based on the yaw rate. The idea is to first estimate the driver intended yaw rate motion and compare it with current yaw rate measurement. If the difference between the intended and actual yaw rate exceeds some threshold, it is decided that an extra yaw moment is necessary in the hope to reduce this yaw rate difference. One way to obtain the intended yaw motion is to relate it to the car steering angle. Take the first two states, the vehicle sideslip angle β and yaw rate r, out of the lateral dynamic equations for the bicycle model in (7.60) and assume fixed tire cornering stiffnesses to obtain:

$$\begin{bmatrix} \dot{\beta} \\ \dot{r} \end{bmatrix} = \begin{bmatrix} -\frac{C_{af}+C_{ar}}{mv_x} & -1 - \frac{C_{af}l_f - C_{ar}l_r}{mv_x^2} \\ -\frac{C_{af}l_f - C_{ar}l_r}{I_z} & -\frac{C_{af}l_f^2 + C_{ar}l_r^2}{I_z v_x} \end{bmatrix} \begin{bmatrix} \beta \\ r \end{bmatrix} + \begin{bmatrix} \frac{C_{af}}{mv_x} \\ \frac{C_{af}l_f}{I_z} \end{bmatrix} \delta_f. \tag{7.89}$$

By setting the state derivates to zeros in (7.89), the steady-state relationship between the yaw rate ω and the front steering δ_f are obtained as

$$\frac{r}{\delta_f} = \frac{v_x}{L + K_{us}v_x^2}, \tag{7.90}$$

where K_{us} is the understeer gradient that was defined earlier in (7.69) and is a constant dependent on wheelbase and cornering stiffness. It is reasonable, then, to use this steady-state yaw rate at a given front wheel steering value via Eq. (7.90) as an approximation of the driver or autonomous driving intended yaw rate.

However, the intended yaw rate estimated from Eq. (7.90) is not always achievable or feasible. The lateral acceleration is limited by the road adhesion limit as

$$|a_y| < \mu g. \tag{7.91}$$

The lateral acceleration can be approximated by

$$a_y = \dot{v}_y + v_x r \approx v_x r, \tag{7.92}$$

if the change of vehicle lateral velocity v_y is not significantly large. Substitution of (7.92) into (7.91) reveals the approximate achievable limits for the yaw rate:

$$|r| < \frac{\mu g}{v_x}. \tag{7.93}$$

Combining the steady-state relationship (7.90) with respect to front-wheel steering and the yaw rate limits (7.93), the intended yaw rate can be defined as follows:

$$r_{ref} = \min \left\{ \frac{v_x}{L + k_{us}v_x^2} |\delta_f|, \frac{\mu g}{v_x} \right\} \cdot \text{sgn}(\delta_f). \tag{7.94}$$

The comparison is then made between the intended yaw rate r_{ref} and the actual yaw rate measured from the on-board rate gyro, with their difference referred to as follows:

$$\Delta r = r_{ref} - r. \tag{7.95}$$

This yaw rate difference Δr is then used to derive how much extra direct yaw moment ΔM is needed. To avoid unnecessary frequent braking actions, the extra yaw moment ΔM is only requested when the magnitude of Δr exceeds some preset threshold $\overline{\Delta r}$, as stated below

$$\Delta M(t) = \begin{cases} 0, & |\Delta r| \le \overline{\Delta r} \\ K_p \cdot \Delta r(t) + K_i \cdot \int_0^t \Delta r(t)dt + K_d \cdot \Delta \dot{r}(t), & |\Delta r| > \overline{\Delta r} \end{cases}. \tag{7.96}$$

using the PID controller in (7.96) to determine the extra yaw moment that is required whenever the error magnitude $|\Delta r|$ is greater than the threshold $\overline{\Delta r}$. The way of delivering this requested extra yaw moment to the vehicle is also considered and is covered in Section 7.5.2, right after we introduce an alternative method for extra yaw moment calculation.

Sideslip Angle-Based Stability Control

The previous method of deriving extra yaw moment based on the difference between the desired yaw rate and the measured one is intuitive and, therefore, straightforward to understand. Another alternative method is less intuitive as it is based on the observation of the vehicle sideslip angle. It is found among studies that the vehicle sideslip angle could be a good reflection of vehicle stability [51, 53]. Similar to (7.90), it is noted that the vehicle sideslip angle also has a similar steady-state relation with the front steering, which could be derived from (7.89):

$$\frac{\beta}{\delta_f} = \frac{l_r - \frac{ml_f v_x^2}{LC_{ar}}}{L + K_{us}v_x^2}. \tag{7.97}$$

Given a certain front-wheel steering angle, if the vehicle manages to maintain its stability, its sideslip angle would eventually die down toward the fixed value approximated by Eq. (7.97). Otherwise, the sideslip angle responds by quickly deviating away from its expected steady-state value and the vehicle motion trend is seen to be that of losing yaw stability, in which case extra yaw moment may be necessary for the vehicle to regain its stability.

A phase plane of the vehicle sideslip angle is hence likely to reveal more clearly the state transition of the sideslip angle and its rate. The lateral and yaw movement of the vehicle can be approximated by the following differential equations based on the bicycle model:

$$m\dot{v}_y = F_{yf} \cos \delta_f + F_{yr} - mv_x r, \tag{7.98}$$

$$I_z \dot{r} = F_{yf}l_f \cos \delta_f - F_{yr}l_r, \tag{7.99}$$

with the assumption that the effect from the longitudinal force to the lateral and yaw dynamics is negligible. The tire lateral forces F_{yf}, F_{yr} can be described by different tire models but generally have the following forms:

$$F_{yf} = f_{tire}(\mu, F_{zf}, \alpha_f, s_f, \ldots), \tag{7.100}$$

$$F_{yr} = f_{tire}(\mu, F_{zr}, \alpha_r, s_r, \ldots), \tag{7.101}$$

which depends on the tire slip angles α_f and α_r, the tire load F_{zf} and F_{zr}, the road friction coefficient μ, and other parameters.

To observe the state transition more accurately, a nonlinear tire model is used to reflect the tire characteristics and its saturation region. The Dugoff tire model used in this case describes the longitudinal and lateral forces for the ith wheel as follows:

$$F_{xi} = f_i C_{xi} s_i,$$ (7.102)

$$F_{yi} = f_i C_{\alpha i} \alpha_i,$$ (7.103)

where the subscript $i = f, r$ represents either front or rear tires for the bicycle model, C_{xi} and $C_{\alpha i}$ are the tire longitudinal and lateral stiffness. s_i is the tire slip ratio as defined to measure the relative difference between the tire's circumferential velocity and its linear velocity referenced to ground and given by

$$s_x = \frac{R_{eff}\omega - v_{tire}}{\max\{R_{eff}\omega, v_{tire}\}}.$$ (7.104)

α_i is the tire sideslip angle for the front or rear tire and can be obtained by

$$\alpha_f = \frac{v_y + l_f r}{v_x} - \delta_f,$$ (7.105)

$$\alpha_r = \frac{v_y - l_r r}{v_x}.$$ (7.106)

Lastly, the coefficient f_i in (7.102) and (7.103) is determined by

$$f_i = \begin{cases} 1 & F_{Ri} \le \frac{\mu F_{zi}}{2} \\ \left(2 - \frac{\mu F_{zi}}{2F_{Ri}}\right) \cdot \frac{\mu F_{zi}}{2F_{Ri}}, & F_{Ri} > \frac{\mu F_{zi}}{2} \end{cases},$$ (7.107)

where

$$F_{Ri} = \sqrt{(c_{xi}s_i)^2 + (c_{yi}\alpha_i)^2}.$$ (7.108)

To obtain state traces in the phase plane of $(\beta, \dot{\beta})$, different value combinations are first set as the initial values for the two initial states in the differential equations (7.98) and (7.99), i.e. lateral speed v_y and yaw rate r. For each set of initial states, these ordinary differential equations (7.98) and (7.99) are then solved with numerical techniques such as the Runge–Kutta method to derive the state traces. Parameters of interest in the equation include the vehicle speed v_x, road friction coefficient μ, and front wheel steering δ_f. For the sake of simplicity, we first start with investigation on the effects of the former two parameters on the phase portrait pattern of $(\beta, \dot{\beta})$. Figure 7.31 shows the state traces in the phase plane of $(\beta, \dot{\beta})$ at different combinations of vehicle speeds and road friction coefficients while the front steering was set to zero. It is revealed that both parameters have an impact on shaping the pattern of the phase portrait. At zero steering wheel angle, the steady-state value of the vehicle sideslip angle naturally lies at the origin. The phase portraits in Figure 7.31 show how the surrounding states either converge to or deviate further away from the steady-state value. It is observed that an increase in speed v_x or a reduction of road friction coefficient μ tend to intensify the trend of the state drifting away.

It is based on the obtained phase portraits that the control law is designed. The region of fast convergence to more stable locations can be confined by two straight lines, with the confined region represented as follows:

$$|\beta + c_1 \dot{\beta}| \le c_2,$$ (7.109)

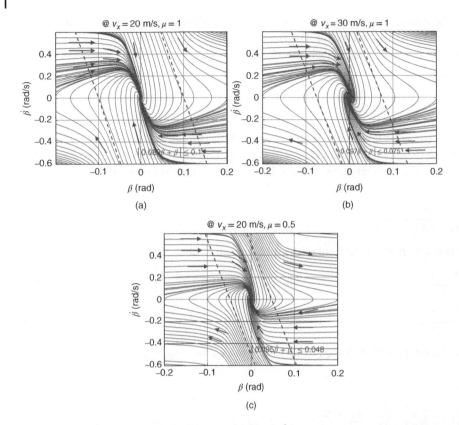

Figure 7.31 Phase portraits of vehicle sideslip angle $(\beta, \dot{\beta})$ as well as the determined control thresholds (in dash) at zero steering and other different parameters: (a) $v_x = 20\,\text{m/s}$, $\mu = 1$; (b) $v_x = 30\,\text{m/s}$, $\mu = 1$; (c) $v_x = 20\,\text{m/s}$, $\mu = 0.5$. Source: Zhu. [54].

where c_1 and c_2 are parameters selected under each operating condition. Examples of these region selections are shown in Figure 7.31, with the arrows indicating the state flow directions. The straight boundary lines in dash along with their coefficient values are selected such that the state within would converge to the equilibrium point quickly. These boundary lines serve as control thresholds that can be used to determine whether extra yaw moment is necessary. If the state is within the region enclosed by (7.109), the vehicle would reach its steady state quickly and hence it is considered safe with no Direct Yaw-moment Control (DYC) needed. However, if the state $(\beta, \dot{\beta})$ is out of this confined region, expressed as

$$|\beta + c_1 \dot{\beta}| > c_2. \tag{7.110}$$

Then extra yaw moment would be added to avoid the state drifting further away from the stationary point or to accelerate the state convergence. One way of determining how much extra yaw moment should be added is to relate it with how far the state reaches beyond the boundary lines. As illustrated in Figure 7.32, the distance of the state from the boundary lines $\Delta\beta$ has the following relationship:

$$\Delta\beta = d \sin\theta, \tag{7.111}$$

Figure 7.32 Distance to the boundary line.

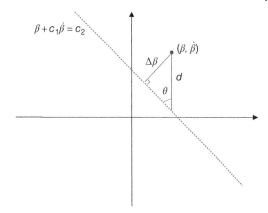

where

$$d = |\beta + c_1\dot{\beta}| - c_2, \tag{7.112}$$

$$\sin\theta = \frac{c_1}{\sqrt{c_1^2 + 1}}, \tag{7.113}$$

The sideslip angle is available from (7.104), and its rate can be obtained by relating the lateral acceleration and yaw rate measurement using

$$\dot{\beta} = \frac{a_y}{v_x} - r. \tag{7.114}$$

Similar to (7.96), a controller such as proportional feedback K_β and sliding action can be used to determine the extra yaw moment based on the deviation $\Delta\beta$ as

$$\Delta M = \begin{cases} 0, & \Delta\beta \leq 0 \\ -K_\beta\Delta\beta \cdot sign\{\beta + c_1\dot{\beta}\}, & \Delta\beta > 0 \end{cases}. \tag{7.115}$$

Following this idea, phase portraits of the sideslip angle at different combinations of road friction coefficient and speeds can be formed, with corresponding control boundaries selected for each condition. The chosen line parameters c_1 and c_2 can, then, be arranged to form a two-dimensional lookup table with respect to the speed and road friction coefficient. It is noted that another dimension, the front steering δ_f, can be added for the stability region selection in the phase portrait. Figure 7.33 shows the phase portrait at the same speed and road friction coefficient as Figure 7.31a except at a 90° steering wheel angle. The comparison shows that the steering not only shifts the stationary point away from the origin but also affects the pattern in the phase plane. Adding this steering angle into consideration can potentially enable a finer adjustment of extra DYC on the vehicle but will require much extra work of parameter tuning as well as having to deal with the dimension and size of the lookup tables. Since the stationary point is usually near the origin, it is reasonable to use only the two parameters of speed and road friction coefficient for extra yaw moment determination.

7.5.2 Direct Yaw Moment Distribution via Differential Braking

Section 7.4.1 introduced the difference between the yaw-rate or sideslip-angle based methods in determining the necessary amount of extra yaw moment. The requested extra yaw moment could

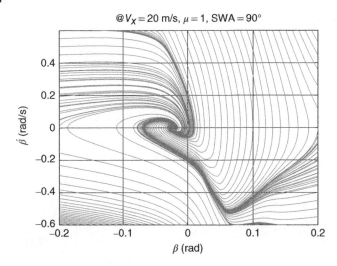

Figure 7.33 Phase portraits of vehicle sideslip angle and its rate $(\beta, \dot{\beta})$ at 90° steering wheel angle. Source: Zhu. [54].

be implemented in limited ways since it eventually relies on the changing of tire forces. As it was mentioned before, the vehicle steerability vanishes as the tire sideslip angle increases in which case providing extra yaw moment through steering compensation is not practical. Then, it is natural to consider utilizing the longitudinal forces from individual wheels to provide this extra yaw moment.

Depending on the vehicle configuration, the slip ratio of the individual tire may be adjusted through the active drive torque and braking torque control for all-wheel-driven vehicles. For most vehicles though, the realization through differential braking is more common and practical. Depending on the directions of the front-wheel steering and the requested yaw moment, the braking force distribution shown in Figure 7.34 can be used. In each case of Figure 7.34, the braking force is applied to a single wheel.

To provide the desired amount of extra yaw moment, the torque required at the braking wheel is

$$T_{br} = \frac{|\Delta M| \cdot r_w}{W/2}, \tag{7.116}$$

where r_w is the tire effective radius and W is the track width. For a disc brake configuration, this would require the amount of clamp load to become

$$F_c = \frac{T_{br}}{r_c \mu_f n}, \tag{7.117}$$

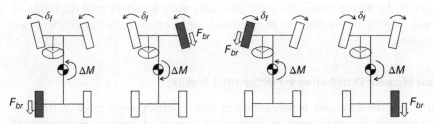

Figure 7.34 Braking force distribution with respect to the direction of front steering and extra yaw moment. Source: Zhu. [54].

where r_c is the effective radius where the brake pads act upon the brake rotor, μ_f is the friction coefficient between the lining material and the disc material, and n is the number of friction surfaces.

Using Eqs. (7.116) and (7.117), the requested wheel cylinder pressure is estimated as

$$p_{req} = \frac{F_c}{A_p} = \frac{2|\Delta M| \cdot r_w}{A_p r_c \mu_f n W}, \tag{7.118}$$

where A_p is the area where the wheel cylinder pressure act upon on the piston to engage the brake pads with the rotor. For the disc brakes of our vehicle, the brake parameter values used in the simulations are $n = 2$, $r_c = 0.13\,\text{m}$, $\mu_f = 0.4$, A_p of the front and rear brakes are 0.0024 and $9.6 \times 10^{-4}\,\text{m}^2$, respectively.

7.5.3 Simulation Results and Discussion

The proposed differential braking control strategies are combined with the front steering control of path tracking for autonomous driving in this subsection. The MPC controller proposed in Section 7.2 is adopted here for steering control, along with either the yaw-rate or sideslip angle-based differential braking control, to observe its path-tracking performance as well as its yaw stability regulation. For simplicity, proportional feedback is selected for both the yaw-rate and sideslip angle-based strategies to determine the amount of extra yaw moment required, with the proportional gains tuned as $k_p = 1 \times 10^4$ in (7.96) and $k_\beta = 2 \times 10^4$ in (7.115), respectively. The DYC via differential braking updates at 100 Hz frequency.

We are interested in potential stability improvements with the proposed DYC strategies when path tracking on a low friction road. A DLC reference path is used to evaluate vehicle performance on this low friction road with $\mu = 0.5$ and speed of 20 m/s under three different controls: using the MPC-based steering control alone, and steering control in combination with the yaw-rate or sideslip-angle based DYC. Figure 7.35 shows the comparisons of their trajectories and the reference path. Figure 7.36 shows the comparisons of vehicle yaw rate and lateral acceleration during the process. Under low μ, the vehicle fails to maintain yaw stability at 20 m/s with the front steering alone. The yaw-rate based and the sideslip-angle based DYC are both capable of ensuring vehicle stability while also maintaining tracking performance.

It is noticeable that the yaw-rate–based control performs worse than the sideslip-angle based one during the low-friction condition simulations, with tracking error up to 2 m in Figure 7.35, and more perturbations in yaw rate and lateral acceleration in Figure 7.36. Its trajectory and state

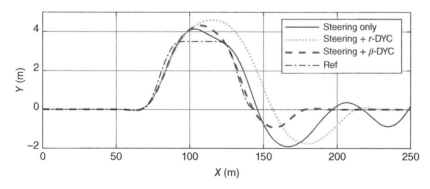

Figure 7.35 Path tracking results with front steering alone versus combined control with DYC at speed 20 m/s and $\mu = 0.5$. Source: Zhu. [54].

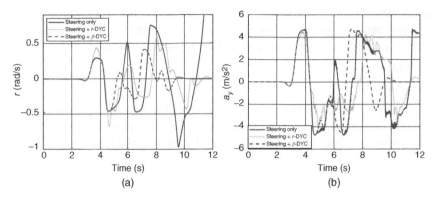

Figure 7.36 Comparison of vehicle yaw rate (a) and lateral acceleration (b) changes at speed 20 m/s and $\mu = 0.5$. Source: Zhu. [54].

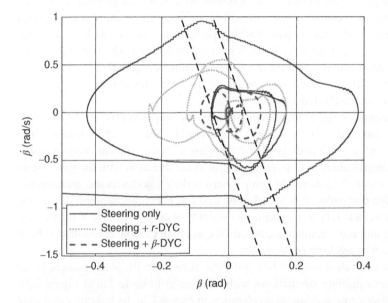

Figure 7.37 Comparison of phase portraits at speed 20 m/s and $\mu = 0.5$. Source: Zhu. [54].

change tend to lag behind others in both plots. This is due to use of the steady-state relationship in Eq. (7.90) that does not model the actual intended yaw rate very well during the transient response period. The estimation may be further distorted by the road adhesion limit in (7.94) for the low μ road. As shown in the phase portrait of Figure 7.37, the sideslip-angle based DYC is able to contain its traces within a smaller region and achieving faster stabilization while, in contrast, the yaw-rate based DYC drifted much further away from the stationary point during the transient response period.

Figures 7.38 and 7.39 show the comparisons of the requested extra yaw moment and individual wheel braking pressures, respectively, for the two DYC strategies. It is observed that the yaw-rate-based control requests more frequent extra yaw moment and correspondingly more individual wheel braking effort. Most of these efforts however are probably unnecessary. On the contrary, the sideslip-angle-based DYC only acts when necessary with far less differential braking

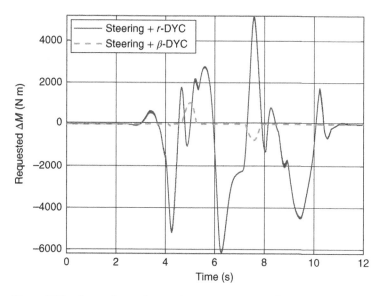

Figure 7.38 Comparison of request extra yaw moment from yaw-rate and sideslip-angle based stability control at 20 m/s, $\mu = 0.5$. Source: Zhu. [54].

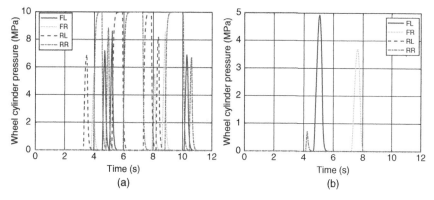

Figure 7.39 Wheel cylinder pressures for individual brake for the (a) yaw-rate based, and (b) sideslip-angle based DYC, at speed 20 m/s, $\mu = 0.5$ ("FL" – front-left, "FR" – front-right, "RL" – rear-left, "RR" – rear-right).

effort, as shown in Figures 7.38 and 7.39, and yet achieves better performance in both stability maintenance and tracking accuracy.

In summary, two different strategies, the yaw-rate and the sideslip-angle-based DYC for determining the necessary amount of extra yaw moment were discussed and compared within an autonomous vehicle path-tracking control system. Both approaches were seen to help the vehicle improve yaw stability and maintain tracking performance while operating on a low friction road. The yaw-rate based control exerted unnecessarily frequent individual braking actuations and was poorer in performance due to estimation discrepancy of the intended yaw rate. The sideslip-angle-based DYC had more favorable results as it was capable of keeping the vehicle state of sideslip within a small region in the phase plane and required very few individual wheel braking interventions.

7.6   Chapter Summary and Concluding Remarks

This chapter summarized different methods for path-tracking control. We specifically covered the design methodology of MPC and robust gain-scheduling control in observance of uncertainty of operating condition such as road, weight, and speed. Real-time implementations of the controllers were provided with comparison and discussions. Moreover, we demonstrated that an add-on VSC can further improve path-tracking accuracy and ensure driving safety.

References

1 X. Xiang, C. Yu, and Q. Zhang, "Robust fuzzy 3D path following for autonomous underwater vehicle subject to uncertainties," *Computers & Operations Research*, vol. 84, no. C, pp. 165–177, 2017.

2 Y. Chen and J. Wang, "Personalized vehicle path following based on robust gain-scheduling control in lane-changing and left-turning maneuvers," in *2018 Annual American Control Conference (ACC)*, 2018, pp. 4784–4789: IEEE.

3 J. Guo, Y. Luo, and K. Li, "Robust gain-scheduling automatic steering control of unmanned ground vehicles under velocity-varying motion," *Vehicle System Dynamics*, vol. 57, no. 4, pp. 595–616, 2019.

4 S. Zhu, S. Y. Gelbal, B. Aksun-Guvenc, and L. Guvenc, "Parameter-space based robust gain-scheduling design of automated vehicle lateral control," *IEEE Transactions on Vehicular Technology*, vol. 68, no. 10, pp. 9660–9671, 2019.

5 G. C. Rains, A. G. Faircloth, C. Thai, and R. L. Raper, "Evaluation of a simple pure pursuit path-following algorithm for an autonomous, articulated-steer vehicle," *Applied Engineering in Agriculture*, vol. 30, no. 3, pp. 367–374, 2014.

6 Y. Chen and J. J. Zhu, "Pure pursuit guidance for car-like ground vehicle trajectory tracking," in *Dynamic Systems and Control Conference*, 2017, vol. 58288, p. V002T21A015: American Society of Mechanical Engineers.

7 M. Elbanhawi, M. Simic, and R. Jazar, "Randomized bidirectional B-spline parameterization motion planning," *IEEE Transactions on Intelligent Transportation Systems*, vol. 17, no. 2, pp. 406–419, 2015.

8 S. Zhu, and B. Aksun-Guvenc, "Trajectory planning of autonomous vehicles based on parameterized control optimization in dynamic on-road environments," *Journal of Intelligent & Robotic Systems*, vol. 100, pp. 1055–1067, 2020.

9 R. S. Sharp, "Driver steering control and a new perspective on car handling qualities," *Proceedings of the Institution of Mechanical Engineers, Part C: Journal of Mechanical Engineering Science*, vol. 219, no. 10, pp. 1041–1051, 2005.

10 S. Zhu, S. Y. Gelbal, X. Li, M. R. Cantas, B. Aksun-Guvenc, and L. Guvenc, "Parameter space and model regulation based robust, scalable and replicable lateral control design for autonomous vehicles," in *2018 IEEE Conference on Decision and Control (CDC)*, 2018, pp. 6963–6969: IEEE.

11 R. E. Kalman, "A new approach to linear filtering and prediction problems," *Journal of Basic Engineering*, vol. 82, no. 1, pp. 35–45, 1960.

12 S. Thrun, W. Burgard, and D. Fox, *Probabilistic Robotics*, MIT Press Cambridge, 2000.

13 S. Quinlan and O. Khatib, "Elastic bands: connecting path planning and control," in *Proceedings IEEE International Conference on Robotics and Automation*, 1993, pp. 802–807: IEEE.

14 S. Y. Gelbal, S. Arslan, H. Wang, B. Aksun-Guvenc, and L. Guvenc, "Elastic band based pedestrian collision avoidance using V2X communication," in *2017 IEEE Intelligent Vehicles Symposium (IV)*, 2017, pp. 270–276: IEEE.

15 S. Y. Gelbal, B. Aksun-Guvenc, and L. Guvenc, "Collision avoidance of low speed autonomous shuttles with pedestrians," *International Journal of Automotive Technology*, vol. 21, no. 4, pp. 903–917, 2020.

16 J. M. Larrazabal and M. S. Peñas, "Intelligent rudder control of an unmanned surface vessel," *Expert Systems with Applications*, vol. 55, pp. 106–117, 2016.

17 Y. Li, X. Zhang, J. Bao, and M. Skyllas-Kazacos, "Control of electrolyte flow rate for the vanadium redox flow battery by gain scheduling," *Journal of Energy Storage*, vol. 14, no. Part 1, pp. 125–133, 2017.

18 A. Moutinho, J. R. Azinheira, E. C. de Paiva, and S. S. Bueno, "Airship robust path-tracking: a tutorial on airship modelling and gain-scheduling control design," *Control Engineering Practice*, vol. 50, pp. 22–36, 2016.

19 D. Saussié, L. Saydy, O. Akhrif, and C. Bérard, "Gain scheduling with guardian maps for longitudinal flight control," *Journal of Guidance, Control, and Dynamics*, vol. 34, no. 4, pp. 1045–1059, 2011.

20 F. Lescher, J. Y. Zhao, J. Yun, "Robust gain scheduling controller for pitch regulated variable speed wind turbine," *Studies in Informatics and Control*, vol. 14, no. 4, p. 299, 2006.

21 R. Tóth, *Modeling and Identification of Linear Parameter-Varying Systems*, Springer, 2010.

22 P. Apkarian and R. J. Adams, "Advanced gain-scheduling techniques for uncertain systems," in *Advances in Linear Matrix Inequality Methods in Control* 2000, pp. 209–228: SIAM.

23 S. Chumalee and J. F. Whidborne, "Gain-scheduled H_∞ control via parameter-dependent Lyapunov functions," *International Journal of Systems Science*, vol. 46, no. 1, pp. 125–138, 2013.

24 M. Sato, "Gain-scheduled output-feedback controllers depending solely on scheduling parameters via parameter-dependent Lyapunov functions," *Automatica*, vol. 47, no. 12, pp. 2786–2790, 2011.

25 F. Wu and K. Dong, "Gain-scheduling control of LFT systems using parameter-dependent Lyapunov functions," *Automatica*, vol. 42, no. 1, pp. 39–50, 2006.

26 B. Zhou, Q. Wang, Z. Lin, and G.-R. Duan, "Gain scheduled control of linear systems subject to actuator saturation with application to spacecraft rendezvous," *IEEE Transactions on Control Systems Technology*, vol. 22, no. 5, pp. 2031–2038, 2014.

27 N. Wada and M. Saeki, "An LMI based scheduling algorithm for constrained stabilization problems," *Systems & Control Letters*, vol. 57, no. 3, pp. 255–261, 2008.

28 F. Blanchini, D. Casagrande, S. Miani, and U. Viaro, "Stable LPV realization of parametric transfer functions and its application to gain-scheduling control design," *IEEE Transactions on Automatic Control*, vol. 55, no. 10, pp. 2271–2281, 2010.

29 J. Mohammadpour and C. W. Scherer, *Control of Linear Parameter Varying Systems with Applications*, Springer Science & Business Media, 2012.

30 H. Zhang and J. Wang, "Active steering actuator fault detection for an automatically-steered electric ground vehicle," *IEEE Transactions on Vehicular Technology*, vol. 66, no. 5, pp. 3685–3702, 2016.

31 H. Zhang, X. Zhang, and J. Wang, "Robust gain-scheduling energy-to-peak control of vehicle lateral dynamics stabilisation," *Vehicle System Dynamics*, vol. 52, no. 3, pp. 309–340, 2014.

32 E. D. Andersen and K. D. Andersen, "The MOSEK interior point optimizer for linear programming: an implementation of the homogeneous algorithm," in *High Performance Optimization*, 2000, pp. 197–232: Springer.

33 F. Schrödel, M. Essam, and D. Abel, "Parameter space approach based state feedback control of LTV systems," in *22nd Mediterranean Conference on Control and Automation*, 2014, pp. 1559–1565: IEEE.

34 B. Demirel and L. Guvenc, "Parameter space design of repetitive controllers for satisfying a robust performance requirement," *IEEE Transactions on Automatic Control*, vol. 55, no. 8, pp. 1893–1899, 2010.

35 A. Peer, T. Schauß, N. Stefanov, *et al.*, "Development of a multi-modal multi-user telepresence and teleaction system," *International Journal of Robotics Research*, vol. 29, no. 10, pp. 1298–1316, 2010.

36 J. E. Hernández-Díez, S.-I. Niculescu, C.-F. Méndez-Barrios, E. J. González-Galván, and R. Hernández-Molinar, "A transparent bilateral control scheme for a local teleoperation system using proportional-delayed controllers," in *2016 12th IEEE International Conference on Control and Automation (ICCA)*, 2016, pp. 473–478: IEEE.

37 H. Wang, A. Tota, B. Aksun-Guvenc, and L. Guvenc, "Real time implementation of socially acceptable collision avoidance of a low speed autonomous shuttle using the elastic band method," *Mechatronics*, vol. 50, pp. 341–355, 2018.

38 M. Walter, N. Nitzsche, D. Odenthal, and S. Müller, "Lateral vehicle guidance control for autonomous and cooperative driving," in *2014 European Control Conference (ECC)*, 2014, pp. 2667–2672: IEEE.

39 S. Zhu, S. Y. Gelbal, and B. Aksun-Guvenc, "Online quintic path planning of minimum curvature variation with application in collision avoidance," in *2018 21st International Conference on Intelligent Transportation Systems (ITSC)*, 2018, pp. 669–676: IEEE.

40 J. Ackermann, P. Blue, T. Bünte, L. Guvenc, D. Kaesbauer, M. Kordt, M. Muhler, and D. Odenthal, *Robust Control: The Parameter Space Approach*, Springer Science & Business Media, 2002.

41 S.-H. Hou, "Classroom note: a simple proof of the Leverrier–Faddeev characteristic polynomial algorithm," *SIAM Review*, vol. 40, no. 3, pp. 706–709, 1998.

42 R. Voßwinkel, L. Pyta, F. Schrödel, I. Mutlu, D. Mihailescu-Stoica, and N. Bajcinca, "Performance boundary mapping for continuous and discrete time linear systems," *Automatica*, vol. 107, pp. 272–280, 2019.

43 D. Odenthal and P. Blue, "Mapping of frequency response performance specifications into parameter space," *IFAC Proceedings Volumes*, vol. 33, no. 14, pp. 531–536, 2000.

44 L. Guvenc, B. A. Guvenc, B. Demirel, and M. T. Emirler, *Control of Mechatronic Systems*, Institution of Engineering and Technology, 2017.

45 H. E. Tseng, J. Asgari, D. Hrovat, P. van der Jagt, A. Cherry, and S. Neads, "Evasive manoeuvres with a steering robot," *Vehicle System Dynamics*, vol. 43, no. 3, pp. 199–216, 2006.

46 C. Hu, H. Jing, R. Wang, F. Yan, and M. Chadli, "Robust H_∞ output-feedback control for path following of autonomous ground vehicles," *Mechanical Systems and Signal Processing*, vol. 70–71, pp. 414–427, 2016.

47 C.-S. Chiu, K.-Y. Lian, and P. Liu, "Fuzzy gain scheduling for parallel parking a car-like robot," *IEEE Transactions on Control Systems Technology*, vol. 13, no. 6, pp. 1084–1092, 2005.

48 K. L. R. Talvala and J. C. Gerdes, "Lanekeeping at the limits of handling: stability via Lyapunov functions and a comparison with stability control," in *ASME 2008 Dynamic Systems and Control Conference*, Ann Arbor, Michiga, 2008, vol. Parts A and B, pp. 361–368.

49 K. L. R. Talvala, K. Kritayakirana, and J. C. Gerdes, "Pushing the limits: from lanekeeping to autonomous racing," *Annual Reviews in Control*, vol. 35, no. 1, pp. 137–148, 2011.

50 *ISO/PAS 21448:2019 – Road vehicles – Safety of the intended functionality*, 2019.

51 E. Liebemann, K. Meder, J. Schuh, and G. Nenninger, "Safety and performance enhancement: The Bosch electronic stability control (ESP)," SAE Technical Paper 2004.

52 J. N. Dang, "Statistical analysis of the effectiveness of electronic stability control (ESC) systems-final report," in NHTSA Technical Report DOT HS 810 794, National Highway Traffic Safety Administration DOT HS 810 794 2007.

53 L. Zhang and G. Wu, "Combination of front steering and differential braking control for the path tracking of autonomous vehicle," SAE Technical Paper Series 2016.

54 S. Zhu, "Path Planning and Robust Control of Autonomous Vehicles," *Doctoral dissertation*, Ohio State University, 2020.

8

Summary and Conclusions

This book treated the important problem of path-planning and path-tracking control with collision avoidance, if needed, in autonomous driving. A summary of the results of Chapters 1–7 is presented in this last chapter along with conclusions.

8.1   Summary

Chapters 1–3 form an introduction and the basic knowledge needed in Chapters 4–7 on path-planning and path-tracking control. The introductory Chapter 1 first treated motivation and introduction for autonomous driving and a brief history of automated driving. This was followed by how Advanced Driver Assistance Systems (ADAS) naturally evolved into autonomous driving functions. Brief information on autonomous driving architectures and a brief review of cybersecurity issues and how they relate to autonomous vehicle path planning and tracking were summarized. Chapter 2 was on the basic tire and longitudinal and lateral vehicle dynamics models that are used in the rest of the book. Both independent slip and combined slip tire force models were presented. The nonlinear road load terms were linearized to obtain a simple, linearized longitudinal dynamics model. The nonlinear, linear, and linear-speed varying single-track vehicle models were used to characterize the lateral dynamics. The transfer functions from the steering input and yaw moment disturbance to yaw rate were derived for the linearized model with speed as a parameter. The nonlinear single-track vehicle model was also augmented with the longitudinal dynamics to formulate a more complete vehicle model with coupled longitudinal and lateral dynamics. The steady-state gain characteristic of the linearized single-track model was used to formulate the understeer gradient and to study understeer and oversteer vehicles as well as their critical and characteristic speeds. The path geometry was combined with the single-track model to obtain both nonlinear and linearized versions of the path-tracking model with path curvature entering the model as a disturbance to be rejected. The simple pure pursuit and the more complicated Stanley steering control methods for path tracking were introduced and compared. The linearized path tracking model developed and presented was modified to form a path-tracking model for driving in reverse. Chapter 3 also presented useful background information that was useful in the understanding of the subsequent Chapters 4–7. Different types of simulations that need to be conducted using the vehicle and path-following models developed in Chapter 2 and higher-fidelity models were treated along with the introduction of model-in-the-loop, hardware-in-the-loop (HIL), traffic-in-the-loop, infrastructure-in-the-loop, and vehicle-in-the-loop simulation. The higher-fidelity CarSim or similar modeling used in such

Autonomous Road Vehicle Path Planning and Tracking Control, First Edition.
Levent Güvenç, Bilin Aksun-Güvenç, Sheng Zhu, and Şükrü Yaren Gelbal.
© 2022 The Institute of Electrical and Electronics Engineers, Inc. Published 2022 by John Wiley & Sons, Inc.

simulations were presented and the HIL simulation setup and vehicles used in the experimental results presented in the book were explained. The unified hardware/software architecture used in these vehicles was explained briefly. Creation of virtual environments for realistic generation of autonomous vehicle perception sensor data was presented using the Unity Engine and LGSVL simulator and the Unreal Engine and CARLA simulator. Chapter 3 also presented a detailed method for estimation of unknown and hard to measure parameters using the estimation of effective tire radius and tire-road friction as examples.

Chapters 4 and 5 were on path planning. Chapter 4 first reviewed some of the commonly used path modeling and generation methods available in the literature. The specific path generation methods presented in Chapter 4 were the use of clothoids, Bezier curves, and polynomial splines. Redistribution of waypoints based on route curvature for a route obtained by data collection or by direct extraction from a map was introduced, followed by smoothing of the waypoints data to reduce the scatter inherent in such experimentally collected values. The least squares curve fit algorithm for polynomial spline coefficient determination for a path divided into segments was presented along with algorithms for computation of path-tracking error after the desired path is determined in the form of discrete waypoints or in the form of polynomial splines. Chapter 5 presented a detailed review of the collision-free path planning methods available in the literature. The methods discussed for collision avoidance included the elastic band method, the quintic spline method with minimum curvature variation, as well as the model-based trajectory planning method. The first method that was proposed deformed the path like an elastic band, using virtual forces exerted by the surrounding obstacles. The efficacy of this elastic band approach was demonstrated in both HIL simulations and experiments that handle both static and moving obstacles. In the second method, the spline path was optimized, aiming at minimum curvature variation of the path for smoothness and passenger comfort, with computational efficiency being achieved by using lookup tables. In combination with a geometric method to generate collision-free target points in the grid map, the quintic spline method was seen to be capable of planning a smooth and safe path in realistic simulations. Chapter 5 also covered the model-based trajectory planning method which added another dimension of the longitudinal motion into planning by generating a set of candidate trajectories to choose from. Its performance was demonstrated on a low friction road, combined with the road friction coefficient estimation module from Chapter 3, and evaluated in different complex driving scenarios with multiple objects.

Chapters 6 and 7 were on path-tracking control. Chapter 6 used a model regulation approach to handle uncertainty and disturbances in the path-tracking model. The disturbance observer (DOB) was applied to the loop from steering input to path-tracking error and use cases for several different applications involving treating uncertainty and rejecting disturbances were presented along with simulation results and analysis. Time delay performance of the loop was also discussed. Chapter 7 started with the model predictive control (MPC) method due to its widespread use in automotive control applications and increasing use in path tracking control. The preview feature of MPC made the resulting steering control smooth and relaxed the requirements on the path description. Its performance in path-following control was evaluated through both simulations and vehicle experiments.

A first requirement on the control law for path following is that the controlled system must remain stable for the different possible operating conditions. This is a robustness of stability requirement and includes nominal stability in the case of no uncertainty. Further guarantee of robustness of performance in the presence of defined uncertainty regions was also seen to be highly desirable. This is a robustness of performance requirement which includes nominal performance in the absence of uncertainty. We used a robust gain-scheduling controller design based on the parameter-space approach to achieve the robust stability and performance requirements. The proportional feedback gain and look-ahead distance used are scheduled with respect to speed while sustaining robust stability, mixed sensitivity bound constraint, and satisfaction of relevant performance indices. Realistic HIL simulation with a validated vehicle model and road tests were used to demonstrate the robust performance of the designed gain-scheduling control in the presence of model uncertainty and disturbances. Another general robust gain-scheduling approach, the linear-matrix-inequality (LMI) design was also conducted later for benchmarking purposes and the comparisons were discussed. Integration of path tracking control with vehicle stability control was also treated in Chapter 7.

8.2 Conclusions

The most essential function of an autonomous system is being able to plan a feasible path to its desired destination and to track that path within acceptable error at the speed requirements of the road while safely avoiding obstacles along the path. The literature is abundant with data-driven studies on end-to-end autonomous driving where very simple and often unrealistic vehicle dynamics models are used. Actual vehicles have mass and mass distribution and many nonlinearities starting with the saturation limits and coupling introduced by the tires. Vehicle model parameters such as mass, speed of operation, and tire-road friction coefficient also change drastically during driving. A realistic and validated model has to be used at higher speeds. The discontinuous and unrealistic nature of most data driven end-to-end approaches to autonomous driving will simply not work in practice. This book, therefore, used a vehicle dynamics based approach to planning and tracking of the path of an autonomous vehicle, including evasive maneuvers needed to avoid collision with obstacles like other vehicles or vulnerable road users. HIL simulations and experimental vehicle testing were used to make sure that the proposed algorithms are implementable and work in real time. As the drive toward L4 and L5 autonomous driving vehicles continues, research including fundamental and applied research on path planning, path-tracking control, and collision avoidance maneuvering of autonomous vehicles will also continue at an accelerated pace.

Index

Autonomous Road Vehicle Path Planning and Tracking Control, First Edition.
Levent Güvenç, Bilin Aksun-Güvenç, Sheng Zhu, and Şükrü Yaren Gelbal.
© 2022 The Institute of Electrical and Electronics Engineers, Inc. Published 2022 by John Wiley & Sons, Inc.

Books in the IEEE Press Series on Control Systems Theory and Applications

Series Editor: Maria Domenica DiBenedetto, University of l'Aquila, Italy

The series publishes monographs, edited volumes, and textbooks which are geared for control scientists and engineers, as well as those working in various areas of applied mathematics such as optimization, game theory, and operations.